Electromagnetism for Signal Processing, Spectroscopy and Contemporary Computing

Electromagnetism for Signal Processing, Spectroscopy and Contemporary Computing

Fundamentals and Applications

Khurshed Ahmad Shah
Brijesh Kumbhani
Raul F. Garcia-Sanchez
Prabhakar Misra

CRC Press is an imprint of the
Taylor & Francis Group, an **informa** business

First edition published 2022
by CRC Press
6000 Broken Sound Parkway NW, Suite 300, Boca Raton, FL 33487-2742

and by CRC Press
2 Park Square, Milton Park, Abingdon, Oxon, OX14 4RN

First edition published by CRC Press 2022

CRC Press is an imprint of Taylor & Francis Group, LLC

ISBN: 978-0-367-75423-5 (hbk)
ISBN: 978-1-032-10062-3 (pbk)
ISBN: 978-1-003-21346-8 (ebk)

DOI: 10.1201/9781003213468

Typeset in Times
by SPi Technologies India Pvt Ltd (Straive)

Dedication

Dr. Khurshed Ahmad Shah dedicates this book to his wife, Sabreena Javid, his son, Muthar Khurshid, and daughter, Hoorain Khurshid, for their cooperation and support.

Brijesh Kumbhani dedicates this book to his wife, Dr. Dipti Borad, and his son, Panthil, for their love and support.

Dr. Raul F. Garcia-Sanchez dedicates this book to the loving memory of his grandfather, Pedro Garcia, to Erin Perley, and his parents, Juan Garcia and Maribel Sanchez, for their love and support.

Dr. Prabhakar Misra dedicates this book to the loving memory of his father-in-law, Dr. Devendra Nath Misra, and in fond memory of his younger brother, Dr. Sudhakar Misra. He also dedicates it to his wife, Suneeta, their son, Uday, and daughter, Isha, for their love of reading and unconditional support of his literary and academic pursuits.

Contents

Preface

Electromagnetism is the basic concept to understanding the properties of matter and is often regarded as the difficult part of a course offering in physics, materials science, or engineering. The classical theories of electromagnetism began at the time of Sir Isaac Newton (1643–1727) and were completed by James Clerk Maxwell and Albert Einstein in the early twentieth century. Microscopic explanations began with British physicist J.J. Thomson's discovery of the electron in 1897. For most physical phenomena, it is best to seek a basic explanation first, especially for phenomena at room temperature, or low energy, when quantum effects are small.

Signal processing includes the science behind our digital lives enabling us to communicate and share information. Spectroscopy, on the other hand, studies the absorption and emission of light and interaction of radiation with matter. Spectroscopic analysis has been crucial in the development of fundamental theories in physics, including quantum mechanics, the special and general theories of relativity, and quantum electrodynamics. Spectroscopy, as applied to high-energy collisions, has been a key tool in developing scientific understanding not only of the electromagnetic force, but also of the strong and weak nuclear forces.

This book discusses various aspects of electromagnetism, signal processing, spectroscopy, and computing. Chapters 1 and 2 provide an overview of the principles of electromagnetism. Chapters 3, 4, and 5 discuss signal processing systems for communication in analog and digital domains. Chapters 6 and 7 cover spectroscopy and computing and associated case studies and applications.

Chapter 1, "Fundamentals of Electromagnetism," gives an introduction to the fundamental concepts of electromagnetism. The chapter covers the fundamentals of electric field and electric flux, Gauss's theorem of electrostatics in integral and differential form, electric potential, electric dipole, and electric dipole moment. The chapter also includes the study of dielectrics, capacitors, and capacitance. The expression for energy stored by a capacitor in an electrostatic field is also derived. These topics lead to a better understanding of magnetic fields. Moreover, the Biot-Savart law is discussed with an expression for divergence of magnetic field. Magnetic scalar and vector potentials are also explained with derivation of the magnetic vector potential. Furthermore, a classification of magnetic materials is given in detail, followed by a short summary of the chapter.

Chapter 2,"Electromagnetic Theory," introduces some of the important and basic concepts taught in different educational institutions all around the world. It covers the fundamentals of Maxwell's equations in matter and in integral form. Then these equations are derived in a simplified manner. The chapter discusses in detail electromagnetic waves, equation of continuity, displacement current, and the expression for displacement current. The concepts of Poynting vector and Poynting theorem are explained in detail. Moreover, the chapter covers electromagnetic wave propagation through vacuum, expression for energy density in electromagnetic field, electromagnetic wave propagation through isotropic dielectric medium, and the transverse nature of electromagnetic waves. Furthermore, polarization and its types are also discussed, followed by a short summary of the chapter.

Chapter 3, "Analog Communication Systems," deals with analog communication systems. We discuss various analog modulation schemes like amplitude modulation (AM) and angle modulation. We compare the analog modulation schemes with respect to the resource requirements like bandwidth and the hardware complexity at both the transmitter and the receiver. We also discuss the circuits for modulation and demodulation. Various types of amplitude modulation, such as double sideband-suppressed carrier (DSB-SC), single sideband (SSB), and AM, are discussed with their advantages and disadvantages. A general structure of radio receiver, called a superheterodyne receiver, is also discussed.

In Chapter 4, "Sampling and Analog to Digital Conversion," we look at the importance of digital systems and discuss the analog to digital conversion of the electrical signals. We discuss various processes involved in analog to digital conversion, such as sampling, quantization, and encoding. Various sampling types have been introduced. The analysis of sampled signal in frequency domain is presented, and the methods to regenerate the original analog signal from sampled signal have been discussed. To represent the digital signal over wired connection, line encoding schemes have been discussed briefly.

In Chapter 5, "Digital Communication Systems," various aspects of digital signal transmission over wireless media are covered. We discuss different pulse-based modulation schemes, as well as digital modulation schemes. The vector representation of digitally modulated signal has been discussed using Gram-Schmidt orthogonalization. Furthermore, digitally modulated signals are represented graphically as signal constellation, using constellation diagrams. The effect of noise on constellation diagram is demonstrated, and its effect on the receiver performance has been discussed. The spread spectrum technique and its immunity to noise are also discussed. Finally, the code division multiple access (CDMA) has been discussed as a technique to serve multiple users over common time frequency resources.

Chapter 6, "Electromagnetism and Spectroscopy," introduces electromagnetism and an overview of some of the most elegant and powerful modern spectroscopy techniques in use today in educational institutions, government, and industrial laboratories all around the world. It covers the fundamentals of absorption and emission spectroscopy, UV-VIS and FT-IR spectroscopy, Raman and X-ray spectroscopy, and mass spectrometry. It also discusses in some depth a variety of applied spectroscopy techniques, with special focus on methodology and practice relating to electrical discharges, condensed phase systems, and materials science. Such techniques lead to a better understanding of discharge plasmas, jet-cooled supersonic expansions, laser-induced breakdown of molecules into smaller fragments, nuclear magnetic resonance phenomena, biophysical and biomedical imaging, laser-mediated photodynamic therapy, and the design and development of novel nanomaterials for advanced sensor applications.

Chapter 7, "Computer Modelling and Simulation, Artificial Intelligence, and Quantum Computing," provides an introduction and overview of model development and visualization techniques, along with data preparation, with emphasis on a variety of popular simulation and computational techniques, namely, large-scale atomic/

molecular massively parallel simulator (LAMMPS), density functional theory (DFT), and COMSOL Multiphysics™. It also covers artificial intelligence, machine learning, and quantum computing, with contemporary applications in the areas of Big Data, materials science, quantum machine learning, and self-driving cars.

We hope that this book will help enhance basic knowledge and understanding in the fields of electromagnetism, signal processing, spectroscopy, and contemporary computing.

Acknowledgements

Dr. Khurshed Ahmad Shah thanks Dr. Rabia Hamid, Head Department of Nanotechnology, University of Kashmir, Srinagar, for her warm heartedness, and his students, Shunaid Parvaiz and Muzaffar Ahmad, for their contribution to this joint endeavor. He is also thankful to his family and his parents for the encouragement and support during the preparation of this book.

Dr. Brijesh Kumbhani thanks all his teachers for making him knowledgeable enough to reach his goals. Special mention goes to the encouragements and guidance he received from Prof. Rakhesh Singh Kshetrimayum at the Indian Institute of Technology Guwahati (IIT Guwahati), during his PhD studies. He also extends his appreciation to his friends, parents, wife, son, and siblings for all their love and support. He is also thankful for the colleagues, students, and staff at IIT Ropar.

Dr. Raul F. Garcia-Sanchez acknowledges the hard work of many people who were responsible for his ability to obtain a higher education, Dr. Juan Arratia from Model Institutions for Excellence (MIE), Dr. Kamla Deonauth from Alliances for Graduate Education and the Professoriate (AGEP), and his advisor Dr. Prabhakar Misra.

Prof. Prabhakar Misra acknowledges the hard work and dedication of the legion of undergraduate and graduate students, research associates, and postdoctoral fellows who were members of the Laser Spectroscopy Laboratory in the Department of Physics & Astronomy at Howard University in Washington, DC, and who contributed to the research and case studies described in Chapters 6 and 7 of the book, along with the multiple funding agencies that made the effort possible.

All the authors would like to thank senior editor Gauravjeet Singh Reen, editorial assistant Lakshay Gaba, and the team at CRC Press for their continuous help and support throughout the project.

Dr. Khurshed Ahmad Shah
Dr. Brijesh Kumbhani
Dr. Raul F. Garcia-Sanchez
Dr. Prabhakar Misra

Acknowledgements

Authors

Dr. Khurshed Ahmad Shah is Senior Assistant Professor in the Higher Education Department and is currently associated with the Department of Nanotechnology at the University of Kashmir, Srinagar, Jammu and Kashmir, India. He earned his Ph.D. in Physics from Jamia Millia Islamia, Central University, New Delhi, India, and a Master of Science and Master of Philosophy from the University of Kashmir, Srinagar, India. He has published research papers in national and international refereed journals and conference proceedings. Dr. Shah has co-authored four books including: *Nanotechnology: The Science of Small* and *Nanoscale Electronics Devices and Their Applications*. He presented his research work in many national and international conferences and guided research scholars. He has broad research interests in the areas of synthesis and characterization of 0D, 1D, and 2D materials and their applications, modeling and simulation of nanoscale electronic devices, sensors, and water purification. He is an editorial board member and reviewer of many scientific journals and member of many scientific and academic associations including: American Physical Society (USA), Institute of Physics (London United Kingdom), Materials Research Society (India), Semiconductor Society (India), IEEE (United States), and International Association of Advanced Materials (Sweden). Dr. Shah has successfully handled three major research projects funded by national agencies. He has received many awards which include the Indian National Science Academy (INSA) Visiting Scientist Fellowship (2019), State Innovative Science Teacher Award (2013), Jawaharlal Memorial Fellowship for Doctorial Studies (2006), Young Scientist Fellowship (2010), and three Jawaharlal Nehru Center for Advanced Scientific Research, Bangalore, India, Visiting Scientist Fellowships.

Dr. Brijesh Kumbhani received his Ph.D. degree from the Department of Electronics and Electrical Engineering, at the Indian Institute of Technology, Guwahati, in 2015. He completed his Bachelor's degree in Electronics and Communication Engineering (ECE) from Dharmsinh Desai University (DDU), Nadiad, India, in 2010. Since June 2016, he has been Assistant Professor at IIT, Ropar. He is the recipient of the IETE S. K. Mitra Memorial Award. His research interests are in the areas of MIMO wireless communication, cloud radio access networks (CRAN), and UWB communication systems. He has delivered talks on recent advances in wireless communications at various national level workshops and symposia. He is a senior member of IEEE, USA, and an overseas member of IEICE, Japan.

Dr. Raul F. Garcia-Sanchez is Media Specialist in the Department of Physics & Astronomy at Howard University (HU) in Washington, DC, USA. He earned his Ph.D. degree in Physics from HU in 2016. He also has a B.S. in Computer Science and a minor in Mathematics. His research interests include laser spectroscopy and computer modeling of nanomaterials and the effects of experimental conditions, such as temperature, humidity, toxic gas exposure, on the Raman spectroscopy of metal oxide gas sensors. His computer programming skills allow him to work in the area of artificial intelligence and machine learning, especially as it relates to modeling and simulation related to large databases. He has been involved in interdisciplinary research that combines physics with computer science, chemistry, mathematics, and even social sciences.

Dr. Prabhakar Misra is Professor of Physics and Director of the Laser Spectroscopy Laboratory in the Department of Physics & Astronomy at Howard University in Washington, DC, USA. He earned a Ph.D. in Physics (1986) from The Ohio State University, Columbus, and an M.S. in Physics (1981) from Carnegie Mellon University, Pittsburgh, PA. Earlier, in India, he had studied at the University of Calcutta where he received his B.Sc. degree in Physics (1975) and an M.Sc. degree in Physics (1978). His research expertise encompasses Experimental Atomic & Molecular Physics and Condensed Matter Physics, especially the twin areas of Raman spectroscopy of nanomaterials and laser spectroscopy of free radicals. Dr. Misra serves as Advisor to the Society of Physics Students (SPS) Chapter at Howard University and has advised and mentored more than 35 undergraduate students, 13 Ph.D. students and 7 postdoctoral research associates, who have been part of his research group in the Laser Spectroscopy Laboratory. Prof. Misra has edited three books and is the author/co-author of more than 215 research abstracts, conference proceedings and refereed journal publications. He is a Visiting Scientist at the NASA Goddard Space Flight Center in Greenbelt, Maryland, USA (2010–present), and a research affiliate at START, a Department of Homeland Security (DHS) Center of Excellence at the University of Maryland, College Park, MD, USA (2014–present). He is a past recipient of the NASA Administrator's Fellowship award (1999–2001) and the Fulbright Scholar Award (2004–2005), and he was elected a Fellow of The American Physical Society (APS) in 2015. He is a recipient of the 2018 NASA Robert H. Goddard Team Award for Exceptional Achievement for Science.

1 Fundamentals of Electromagnetism

1.1 INTRODUCTION

Electromagnetic field theory is the result of the union of three distinct sciences. The oldest of these is electrostatics, which was first studied by the Greeks. Electrostatics is the branch of physics which deals with the static electricity, or, in other words, electrostatics deals with the charges at rest. It is also referred to as frictional electricity, because such charge is produced in insulators rubbing each other. Historically, the Greeks knew something about electric forces arising from frictional electric charges. This was as early as in the sixth century B.C. One of the founding fathers of Greek science, Thales of Miletus, realized that when amber is rubbed against wool, it is able to attract pieces of straw. This observation lays the basis for the science of electricity. Elektron is the Greek name for amber, which is the origin of the words electric charge, electric force, electricity, and electron. In 1785 French physicist Charles-Augustin de Coulomb (1736–1806) showed that electrically charged materials sometimes attract and sometimes repel each other. This was the first indication that there were two types of charge: positive and negative. Charge is the inherent property of the matter which causes it to experience force when exposed to an electric field. The electric charge is of two types, carried by electrons (negative charge) and protons (positive charge). The like charges repel each other, while the unlike charges attract each other. The objects which carry the negative charge and positive charge in equal quantity are referred as "neutral".

When an ebonite rod is rubbed against flannel (soft woolen material) or a glass rod is rubbed against silk, they acquire (or gain) a power to attract light bodies such as small pieces of paper. The agency which gives this attracting power is called 'electricity'. The bodies which acquire this power are referred to as being charged or electrified. The electricity that is created via friction is referred to as the frictional electricity. The electricity is referred to as the static electricity if the charges do not move in a body. When a plastic comb is used to comb dry hair, the comb acquires the capability to attract small bits of paper. However, this is true only if there is no humidity in the atmosphere.

While the ancient Greeks knew about magnetism in the form of lodestone, the Chinese invented the magnetic compass, and, in 1600, William Gilbert of Gloucester laid down some fundamentals in a book titled, *De Magnete* (on the magnet), where he gave the first scientific account of the early experiences of electric and magnetic effects. He introduced the name 'electrica' for substances like amber which when rubbed with suitable materials attracted light bodies. He also stated in his book that in all experiments on frictional electricity, two kinds of electricity are produced. He

DOI: 10.1201/9781003213468-1

stated that while electric charges of the similar type repel each other, those of opposite kinds attract one other. The two kinds of electricity were called resinous and vitreous. The type of electricity generated on amber when it is rubbed with wool was called resinous probably because amber is a resin, and the type of electricity produced on wool was called vitreous. However, it was not until 1785 that Charles-Augustin de Coulomb (1736–1806) formulated his law relating the strengths of two magnetic poles to the force between them. Magnetism may have been laid to rest here if it was not for the Danish physicist Hans Christian Oersted (1777–1851). It was Oersted who demonstrated to a group of students that a current-carrying wire produces a magnetic field. This was the first sign that electricity and magnetism could be interlinked.

An American scientist and statesman, Benjamin Franklin (1706–1790) presented a convention where the charge that appears on ebonite or amber when rubbed with wool was named as negative, while the charge that appears on the glass was named as positive. So 'vitreous' became positive, while 'resinous' became negative. However, it is due to the work of the German physicist Heinrich Rudolf Hertz (1857–1894) and the Italian engineer Guglielmo Marconi (1874–1937) that we are now able to communicate over vast distances. We can also use electrical machinery to make our lives more comfortable.

1.2 ELECTRIC FIELD

A charge generates an electric field in its surroundings and is responsible for the other charge particles to observe a force when placed around its vicinity. The electric field exists on its own without the presence of the other test charge to experience the force. Thus, the electric field in simple terms can be expressed as the space or region surrounding a charged body (source charge) where the force of attraction or repulsion (electric influence) can be experienced by another charge particle (test charge).

Let an electric charge $+q$ be situated at point (O) in space as depicted in Figure 1.1, and a unit positive charge ($+q_0$) be located near it (say at P), then $+q_0$ will observe a force of repulsion. For a negative charge $-q$, the charge ($+q_0$) will observe a force of attraction. The electric field of $+q$ or $-q$ at O applies a force on $+\ q_0$ placed at P.

Therefore, the electric field owing to a charge particle refers to the space in the vicinity of the charge where other charge particles observe a force of repulsion or attraction. Furthermore, in this the charge $+q$ is referred to as the 'source charge' since it is responsible for the creation of the electric field, whereas the charge $+q_0$ is referred to as the 'test charge'. The test charge should be a small as possible, so that the source charge does not affect the electric field due to its presence.

The electric field intensity in the electric field at any given point is defined as the force observed by a unit positive charge placed at the same point. Let us consider a

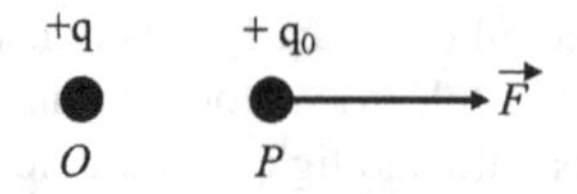

FIGURE 1.1 Electric charge $+q$ situated at point (O) in space.

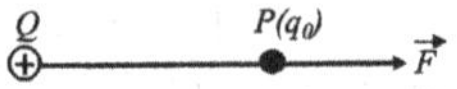

FIGURE 1.2 Diagram showing test charge $+q_0$ located at a point P in the electric field.

test charge $+q_0$ be located at a point P in the electric field due to a source charge Q as depicted in Figure 1.2. Also let $\vec{F}$ be the force exerted by the source charge over the test charge q_0.

Therefore, the electric field intensity at any point P in its vicinity in terms of the force experienced by the test charge due to the source charge is given by

$$\vec{E} = \frac{\vec{F}}{q_0} \tag{1.1}$$

Furthermore, the direction of the electric field intensity is the one in which a unit positive charge will move, if it is let free. Also, the S.I. unit of electric field intensity is Newton/Coulomb (NC^{-1}). Here q_0 is very small, so its presence does not affect the electric field due to the source charge. Thus, the electric field intensity can be more appropriately written as,

$$\vec{E} = \lim_{q_0 \to 0} \frac{\vec{F}}{q_0} \tag{1.2}$$

Additionally, theoretically the electric field extends up to infinity due to the source charge. However, the electric field strength decays quickly with the distance from the source charge. Electric field intensity is vector quantity; it has both the magnitude as well as the direction.

1.3 ELECTRIC FLUX

Electric flux via any surface gives the number of electric field lines crossing through it, such that the electric lines of force are perpendicular to the surface. Consider an arbitrary surface (S) placed in a non-uniform electric field (E). Also consider at a point P an infinitesimal surface with area dS, where the intensity of electric field vector corresponding to the area element $\vec{dS}$ is ($\vec{E}$) as shown in Figure 1.3. The angle between electric field $\vec{E}$ and the area element $\vec{dS}$ at P is (θ). The scalar product $\vec{E} \cdot \vec{dS}$ gives the electric flux through the area element $\vec{dS}$.

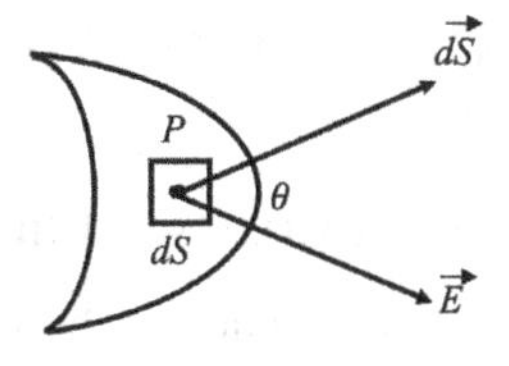

FIGURE 1.3 Diagram showing infinitesimal surface dS with a point P on it.

Consider a hypothetical plane surface of area ΔS and suppose a uniform electric field E exists in the space. Let the electric field E make an angle θ with the positive normal, then the flux of the electric field through the chosen surface is given by,

$$\Delta\Phi = E\Delta S Cos(\theta) \tag{1.3}$$

If we draw a vector of magnitude ΔS along the positive normal, it is called the area vector corresponding to the area. Equation (1.3) can be written as,

$$\Delta\Phi = E \cdot \Delta S \tag{1.4}$$

The electric flux is a scalar quantity and its S.I. unit is Nm^2C-1. Thus, if the surface ΔS has two parts $\Delta S1$ and $\Delta S2$, the flux through ΔS equals the flux through $\Delta S1$ plus the flux through $\Delta S2$. This gives us a clue to define the flux through surfaces which are not plane, as well as the flux when the field is not uniform. We divide the given surface into smaller parts so that each part is approximately plane and the variation of electric field over each part can be neglected. We calculate the flux through each part separately using the relation $\Delta\Phi = E \cdot \Delta S$ and then add the flux through all the parts. Therefore, the electric flux can be calculated via the whole closed surface S by incorporating surface integral given by

$$\phi = \int_S \vec{E} \cdot \vec{d}S \tag{1.5}$$

It is clear that this electric flux is the surface integral over the surface S of the electric field $\vec{E}$. Thus, the electric flux associated with a surface in an electric field can be defined as the surface integral of the electric field over the same surface.

The total flux through the surface, such that the surface S encloses a volume (like the surface of a football) is given by

$$\phi = \oint_S \vec{E} \cdot \vec{d}S \tag{1.6}$$

From the above discussion, we conclude that when electric field lines are perpendicular to the plane of the surface, the electric flux through a given surface is maximum, and when electric field lines are along the plane of the surface, the electric flux through a given surface is zero or minimum.

1.4 GAUSS'S THEOREM OF ELECTROSTATICS IN INTEGRAL AND DIFFERENTIAL FORM

1.4.1 Gauss's Theorem of Electrostatics in Integral Form

The electric flux passing through any imaginary spherical surface lying between the two conducting spheres is equal to the charge enclosed within that imaginary surface.

This enclosed charge is distributed on the surface of the inner sphere, or it might be concentrated as a point charge at the center of the imaginary sphere. However, because one coulomb of electric flux is produced by one coulomb of charge, the inner conductor might just as well have been a cube or a brass door key, and the total induced charge on the outer sphere would still be the same. Gauss's law in integral form states that the total electric flux ϕ through any closed surface is equal to $\frac{1}{\varepsilon_0}$ times the total charge enclosed by that surface. Mathematically, the electric flux ϕ crosses through a closed surface (S) enclosing a charge (q) is given by

$$\phi = \oiint_S \vec{E} \cdot \vec{dS} = \frac{q}{\varepsilon_0} \tag{1.7}$$

And

$$\phi = \oiint_S \vec{E} \cdot \vec{dS} = 4\pi q \tag{1.8}$$

Where $\vec{E}$ = electric field vector and $\vec{dS}$ = vector elements of area pointing in the outward normal direction from the surface. If the charge q lies external to the closed surface (S) or no net charge is enclosed by the surface, then

$$\phi = \oiint_S \vec{E} \cdot \vec{dS} = 0 \tag{1.9}$$

To prove the above statement, let us consider a point charge ($+q$) inside the closed surface (S) as depicted in Figure 1.4. The electric field intensity constituted at an element of area $\vec{dS}$ of the surface, at a distance (r) from q in vacuum, is given by

$$\vec{E} = \frac{1}{4\pi\varepsilon_0} \frac{q}{r^2} \hat{r} \tag{1.10}$$

Where $\hat{r}$ is the unit vector in the direction of $\vec{E}$.

∴ Electric flux through the area $\vec{dS}$ is given by

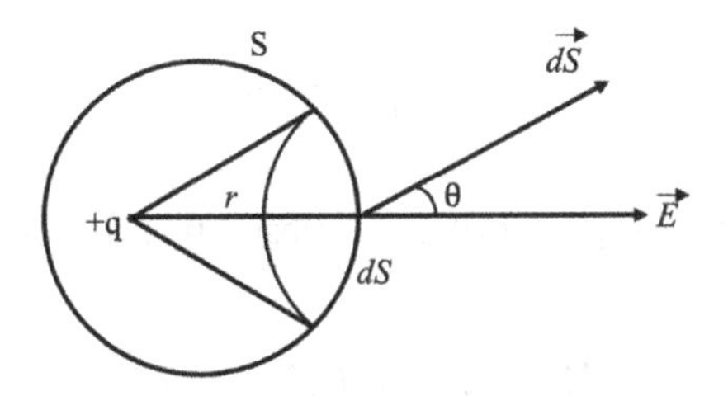

FIGURE 1.4 Diagram showing point charge inside a closed surface.

$$d\phi = \vec{E}\cdot\vec{dS} = |\vec{E}||\vec{dS}|\cos\theta \tag{1.11}$$

where θ is the angle in between $\vec{E}$ and $\vec{dS}$.

Substituting Equation (1.10) in Equation (1.11), we get

$$d\phi = \frac{1}{4\pi\varepsilon_0}\frac{q}{r^2}\left|\vec{dS}\right|\cos\theta$$

or

$$d\phi = \vec{E}\cdot\vec{dS} = \frac{q}{4\pi\varepsilon_0}\cdot d\Omega \tag{1.12}$$

where $d\Omega = \dfrac{\left|\vec{dS}\right|\cos\theta}{r^2}$ is the solid angle subtended by the area element $\vec{dS}$ at the point q.

∴ The net flux over the whole surface can be calculated by integrating Equation (1.12) over the whole surface, i.e.,

$$\phi = \oiint_S \vec{E}\cdot\vec{dS} = \frac{q}{4\pi\varepsilon_0}\oiint_S d\Omega = \frac{q}{4\pi\varepsilon_0}\times 4\pi = \frac{q}{\varepsilon_0}$$

Where, $\oiint_S d\Omega = 4\pi$ is the solid angle

or

$$\phi = \oiint_S \vec{E}\cdot\vec{dS} = \frac{q}{\varepsilon_0} \tag{1.13}$$

which is the required integral form of Gauss's theorem.

In CGS system $\frac{1}{\varepsilon_0} = 4\pi$, ∴ Equation (1.13) becomes

$$\phi = \oiint_S \vec{E}\cdot\vec{dS} = 4\pi q \tag{1.14}$$

1.4.2 Gauss's Theorem of Electrostatics in Differential Form

The charge enclosed by the surface enclosing volume V is given by

$$q = \iiint_V \rho dV$$

Where ρ is the volume charge density. By substituting the value of q in Equation (1.12), we get

$$\oiint \vec{E}\cdot\vec{dS} = \frac{1}{\varepsilon_0}\iiint_V \rho dV$$

or

$$\varepsilon_0 \oiint \vec{E} \cdot \vec{dS} = \iiint_V \rho dV \tag{1.15}$$

Now as per Gauss's divergence theorem

$$\iiint_V \left(\vec{\nabla} \cdot \vec{E}\right) dV = \oiint_S \vec{E} . \vec{dS}$$

or

$$\oiint_S \vec{E} \cdot \vec{dS} = \iiint_V \left(\vec{\nabla} \cdot \vec{E}\right) dV$$

∴ Equation (1.15) becomes

$$\varepsilon_0 \iiint_V \left(\vec{\nabla} \cdot \vec{E}\right) dV = \iiint_V \rho dV$$

Since it satisfies for any arbitrary volume, integrands should be equal

$$\therefore \varepsilon_0 \vec{\nabla} \cdot \vec{E} = \rho$$

or

$$\vec{\nabla} \cdot \vec{E} = \frac{\rho}{\varepsilon_0} \tag{1.16}$$

This represents Gauss's theorem in its differential form.

1.5 ELECTRIC POTENTIAL

1.5.1 Electric Potential as Line Integral of Electric Field

Vector functions like electric or magnetic field appear in physical applications, and the scalar products of such vector functions with other vectors like distance or path length occur with consistency. When such a product is added over a track length, such that the magnitudes and directions change, that addition becomes a line integral. The electric potential is defined as the line integral of electric field in a space between any two points.

Suppose a source charge $+q$ is located at the origin O and a test charge q_0 is held in equilibrium at position P with position vector $\vec{r}$ from origin O. Let $\vec{E}$ represent the electric field at point P due to some charge q over the charge q_0 at P as shown in Figure 1.5.

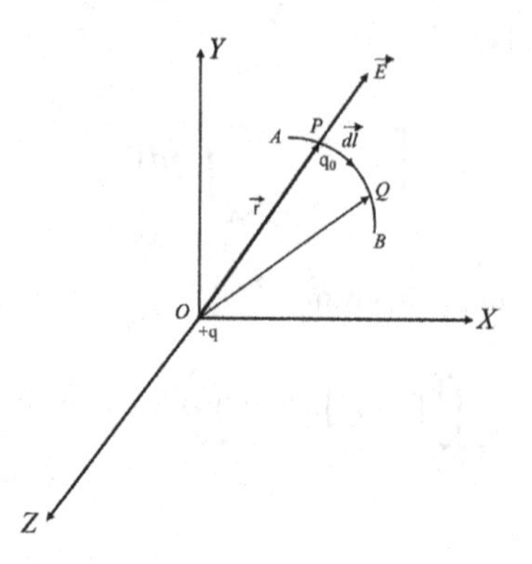

FIGURE 1.5 Diagram showing electric field at point P due to some charge q over the charge q_0 at P.

Therefore, the amount of force needed to keep the charge q_0 at P in equilibrium is given by

$$\vec{F} = -q_0\vec{E}$$

The negative sign depicts that an equal but opposite force is needed to keep the charge q_0 at P in equilibrium.

Let the test charge q_0 be moved by a distance $\vec{dl}$. Then the force does the work given by

$$\mathrm{dW} = \vec{F} \cdot \vec{dl} = -q_0\vec{E} \cdot \vec{dl}$$

To dislocate the charge q_0 from A to B, the total work done is given by

$$W = \int dW = -q_0 \int_A^B \vec{E} \cdot \vec{dl} \tag{1.17}$$

$$\text{If } q_0 = 1, \quad \mathrm{W} = -\int_A^B \vec{E} \cdot \vec{dl} \tag{1.18}$$

In magnitude

$$\mathrm{W} = \int_A^B \vec{E} \cdot \vec{dl} \tag{1.19}$$

Thus the term $\int_A^B \vec{E} \cdot \vec{dl}$ is referred as the line integral of electric field between the two points A and B. The line integral of electric field along a specific path is defined as the amount of work done due to the field in displacing a unit positive charge along that path.

1.5.2 Electric Potential due to a Point Charge

Let us take a point charge $+q$ at origin O and calculate the electric potential owed by this point charge at any point say P at a distance of r from the origin in the electric field $\vec{E}$ due to the point charge $+q$ as shown in Figure 1.6.

As we know, the potential in an electric field $\vec{E}$ at point P, is given by

$$V = \int_{\infty}^{P} \vec{E} \cdot \vec{dl} \tag{1.20}$$

Let the unit test charge be moved from point P to point Q, so that its displacement is equal to $\vec{dr}$

Now $\vec{dl} = -\vec{dr}$

$$\therefore \vec{E} \cdot \vec{dl} = -\vec{E} \cdot \vec{dr} = -Edr\cos 180° = \text{Edr} \tag{1.21}$$

By substituting the value of $\vec{E} \cdot \vec{dl}$ in Equation (1.20), we get

$$V = -\int_{\infty}^{r} Edr \tag{1.22}$$

At point P the magnitude of electric field $\vec{E}$ is given by

$$E = \frac{1}{4\pi\varepsilon_0} \frac{q}{r^2} \tag{1.23}$$

By putting the value of E in Equation (1.22), we get

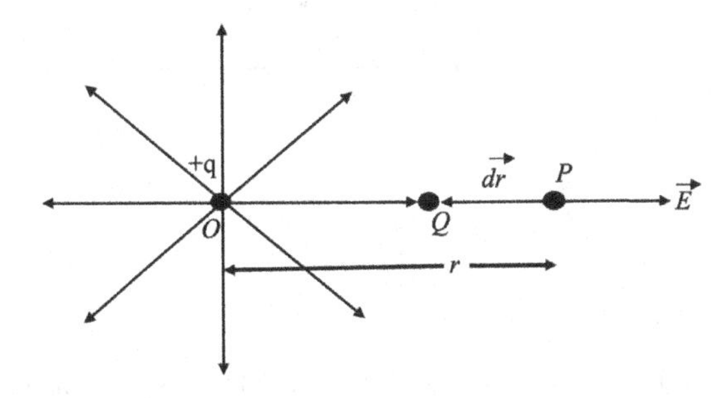

FIGURE 1.6 Diagram showing electric potential due to a point charge.

$$V = -\int_{\infty}^{r} \frac{1}{4\pi\varepsilon_0} \frac{q}{r^2} dr$$

$$\text{or } V = -\frac{q}{4\pi\varepsilon_0} \int_{\infty}^{r} r^{-2} dr$$

$$\text{or } V = -\frac{q}{4\pi\varepsilon_0} \left[-\frac{1}{r} \right]_{\infty}^{r}$$

$$\text{or } V = -\frac{q}{4\pi\varepsilon_0} \left[-\frac{1}{r} + \frac{1}{\infty} \right]$$

$$\text{or } V = \frac{1}{4\pi\varepsilon_0} \frac{q}{r} \tag{1.24}$$

Thus, electric potential via a point charge q at a distance of r is directly proportional to the magnitude of point charge q and inversely proportional to the distance r from the point of observation. Furthermore, let us suppose at the point A, a point charge q from the origin O, Its position vector $\vec{r}$ as depicted in Figure 1.7, Then the electric potential at point P with position vector $\vec{r_1}$ in the electric field due to charge q is given by

$$V_p = \frac{1}{4\pi\varepsilon_0} \frac{q}{\left| \vec{r_1} - \vec{r} \right|}$$

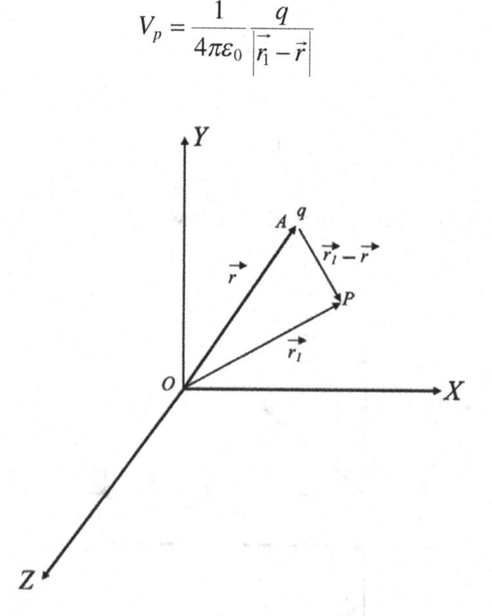

FIGURE 1.7 Diagram showing electric field at point P due to a point charge q.

1.6 ELECTRIC DIPOLE AND ELECTRIC DIPOLE MOMENT

1.6.1 Electric Dipole

An electric dipole is defined as a pair of opposite charges, say q and $-q$, separated by some distance d. By convention, the direction of electric dipoles in space is always from negative charge $-q$ to positive charge q. The midpoint between the two charges q and $-q$ is named as the center of the dipole. Figure 1.8 represents an electric dipole comprising of two charges $-q$ and $+q$ separated by distance of $AB = 2l$. The distance AB is referred as the length of the dipole and is represented by a vector $\vec{2l}$ whose direction is from charge $(-q)$ to charge $(+q)$.

1.6.2 Electric Dipole Moment

The electric dipole moment of a system gives a measure of the separation between the negative and positive charges and is defined as a product of magnitude of either charge and separation between the two charges, mathematically

$$\vec{P} = q \times \vec{2l}$$

The S.I. units of electric dipole moment is coulomb-meter (C · m); however, a commonly used unit in atomic physics and chemistry is the Debye (D) and is denoted by $\vec{P}$. The vector $\vec{l}$ is drawn from the negative to positive charge and is along the axis of the dipole. Furthermore, the direction of electric dipole moment is from the negative to positive charge.

1.7 DIELECTRICS

Dielectric material is an insulating material or a material that has very poor conduction property of electric current. When placed in an electric field, no current flows through because, unlike metals, dielectrics have no free electrons that may drift through the material. Instead, electric polarization occurs. The positive charges within the dielectric are displaced in the direction of the electric field, and the negative charges are displaced in the direction opposite to that of the electric field. This separation of charge, or polarization, decreases the electric field inside the dielectric. The resistivity of dielectric materials is very high (of the order of 10^{22} $\Omega - m$), and examples are glass, mica, paper, air, etc. The dielectric constant (K) gives the measure of the electric potential energy produced in the form of induced polarization under the action of an electric field that is stored in a given volume of material. It is

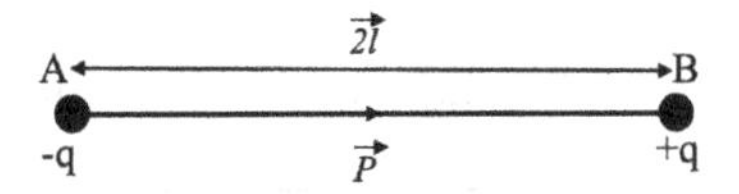

FIGURE 1.8 Diagram representing an electric dipole comprising of two charges $-q$ and $+q$ separated by a distance of $AB = 2l$.

represented as the ratio of the dielectric permittivity of the material to that of a dry air or vacuum.

Consider a rectangular slab of a dielectric material as shown in Figure 1.9. When an external electric field $\overrightarrow{E_0}$ is applied on the slab, +*ve* and −*ve* induced charges grow due to polarization on the dielectric surfaces and create an induced electric field $\overrightarrow{E_P}$, whose direction is opposite to that of $\overrightarrow{E_0}$. This field $\overrightarrow{E_P}$ due to polarization is called the 'depolarizing field'.

For a homogeneous and isotropic dielectric, the polarization is uniform and the resultant field $\vec{E}$ inside the dielectric is given by

$$\vec{E} = \overrightarrow{E_0} - \overrightarrow{E_P} \tag{1.25}$$

where the resultant field $\vec{E}$ is in the direction of the applied field $\overrightarrow{E_0}$. Thus we see that the electric field $\vec{E}$ inside the dielectric is of course in the direction of the field $\overrightarrow{E_0}$, but its magnitude is less than the magnitude $\overrightarrow{E_0}$. For a particular dielectric material, the ratio of the applied electric field $\overrightarrow{E_0}$ and the resultant field $\vec{E}$ inside the dielectric is a constant K, whose value is greater than one (1). This constant (*K*) is called the dielectric constant of the material, i.e.,

$$K = \frac{\overrightarrow{E_0}}{\vec{E}} \text{ or}$$

$$\vec{E} = \frac{\overrightarrow{E_0}}{K} \tag{1.26}$$

In a dielectric material, if a very high electric field is applied, the outer electrons of the atoms may get detached from them. The dielectric then behaves like a conductor and this phenomenon is called 'dielectric breakdown'. The minimum electric field at which the breakdown occurs is known as the dielectric strength of the material.

1.8 CAPACITOR AND CAPACITANCE

1.8.1 Capacitor

The basic function of a capacitor is to store electrical energy. A capacitor is a device that is capable of storing electrical energy and charge in a small space. It consists of

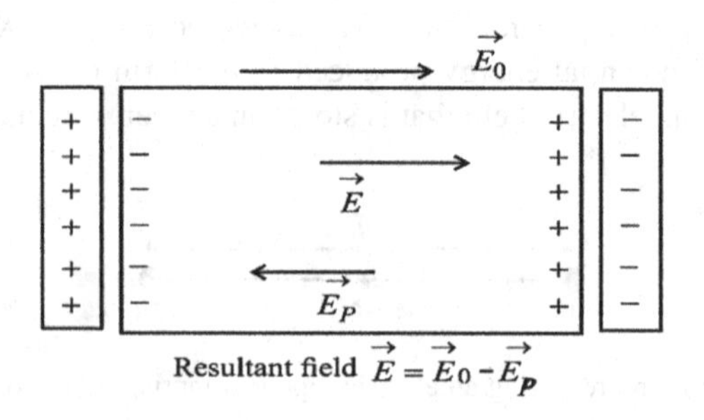

FIGURE 1.9 Diagram showing rectangular slab of dielectric material.

two conducting surfaces sandwiched by an insulating material. The conducting surfaces are termed as 'plates of the capacitor', and the insulating material is named 'dielectric'. The most commonly used dielectrics are air, mica, paper, etc. Furthermore, a capacitor is generally named after the dielectric used, e.g., air capacitor, mica capacitor, paper capacitor, etc. and it may be in the form of a parallel plate (parallel plate capacitor), concentric cylinders (cylindrical capacitor), or any other arrangement.

1.8.2 Capacitance

The capability of a capacitor to store charge is recognized as its capacitance. It is the ratio of the quantity of electric charge stored on a conductor to the electric potential difference. Experimentally, charge (q) stored in a capacitor is directly proportional to potential difference (V) across the plates of the capacitor, i.e.,

$$q \alpha V$$

$$\text{or } q = CV$$

$$\text{or } \frac{q}{V} = C$$

Where C is the constant of proportionality called 'capacitance of the capacitor'. Hence capacitance may be defined as, 'the ratio of charge on capacitor plates to the potential difference across the plates'. We can also define capacitance of a capacitor in another way

$$C = \frac{q}{V}$$

If $V = 1$, then $C = q$. Therefore, capacitance of a capacitor is the amount of charge required to raise its potential by unity. The capacitance of a capacitor depends on shape of its plates, separation between the plates, and nature of insulating material (dielectric) between the plates. The S.I. unit of capacitance is $\frac{Coulomb}{Volt}$, which is also called farad (symbol F), in honor of the famous scientist Michael Faraday. Furthermore, a capacitor is said to have a capacitance of 1 farad if a charge of 1C accumulates on each plate when a potential difference of 1V is applied across the plates.

1.9 EXPRESSION FOR ENERGY STORED BY A CAPACITOR IN AN ELECTROSTATIC FIELD

The process of charging a capacitor is equivalent to the amount of work done in transforming charge from one plate to the other plate of the capacitor. At any instant of charging a capacitor, there is a potential difference between the plates of the capacitor. Therefore, work must be done to transfer the charge from one plate to another plate of the capacitor. This work done is stored as the electrostatic potential energy in the capacitor. Let at any instant the charge on the capacitor be (q) during the charging process, then the potential difference between the plates of the capacitor is given by

$$V = \frac{q}{C} \tag{1.27}$$

where C is the capacitance of the capacitor. If an additional charge, dq, is transferred to the capacitor, then the work done during the process is given by

$$dW = Vdq$$
$$= \frac{q}{C}dq$$

This work done is stored as the electrostatic potential energy in the capacitor, i.e.,

$$dU = dW = \frac{q}{C}dq \tag{1.28}$$

The total electrostatic potential energy while charging the capacitor from $q = 0$ to $q = Q$ is therefore given by

$$U = \int_o^Q \frac{q}{C}dq = \frac{1}{C}\left[\frac{q^2}{2}\right]_o^q$$

or

$$U = \frac{1}{2}\frac{Q^2}{C} \tag{1.29}$$

This energy is stored in the capacitor in the form of electrostatic field energy. Since $Q = CV$

$$\therefore U = \frac{1}{2}CV^2 \tag{1.30}$$

Also $C = \frac{Q}{V}$

$$\therefore U = \frac{1}{2}QV \tag{1.31}$$

The energy stored in the capacitor per unit volume of the space between the plates of the capacitor (i.e., volume occupied by the electric field) is called energy density, i.e.,

Energy density $U_E = \dfrac{U}{Volume}$

For a parallel plate capacitor,

$$\text{Volume} = Ad$$

Therefore, energy per unit volume or energy density

$$U_E = \frac{\frac{1}{2}CV^2}{Ad} \tag{1.32}$$

For parallel plate capacitor $C = \dfrac{A\varepsilon_\circ}{d}$; therefore, Equation (1.32) can be written as

$$\therefore U_E = \frac{\left[\frac{1}{2}\frac{A\varepsilon_\circ}{d}V^2\right]}{Ad}$$

$$\text{or } U_E = \frac{1}{2}\varepsilon_\circ\left[\frac{V}{d}\right]^2$$

the magnitude of electric field , E = V/d

$$\therefore \mathrm{U_E} = \frac{1}{2}\varepsilon_\circ E^2$$

Thus, energy per unit volume (or energy density) in electrostatic field is directly proportional to the square of the magnitude of the electric field.

1.10 MAGNETIC FIELD AND BIOT-SAVART LAW

Magnetostatics is the branch of physics that refers to the study of magnetic fields in a system, so that the currents belonging to the system are steady. It is the analogue of electrostatics. A magnetic field describes the magnetic influence over the moving charge. It was as early as 1802 that an Italian philosopher and jurist named Gian Domenico Romagnosi observed that a magnetic needle is affected by electric current flowing in a wire. This observation was published in a local newspaper, *Gazetta di Trentino*, but was ignored. The connection between electricity and magnetism was rediscovered by a Danish physicist Hans Christian Oersted in 1820; during a lecture demonstration, Oersted observed that a magnetic compass needle aligned itself perpendicular to a current carrying wire. Oersted also noticed that when the direction of the current in the wire was reversed, the direction in which the needle was pointed was also reversed. These observations led Oersted to conclude that a magnetic field is associated with an electric current.

The quantitative consequences of a steady current flow were established by four French physicists. François Arago demonstrated that a current carrying wire behaves like an ordinary magnet in its ability to attract iron fillings. André-Marie Ampère discovered that current carrying wires exert forces of attraction or repulsion on each other. He also determined the laws governing these interactions. Also, Jean-Baptiste Biot and Felix Savart determined experimentally the magnitude and direction of the magnetic field due to small current element. The S.I. unit of a magnetic field $\vec{B}$ is Tesla (T) or weber per square meter (Wb/m^2).

A moving electric charge when placed in a magnetic field practices a force perpendicular to the magnetic field and its own velocity. It is a space around a current carrying conductor (or a magnet) where magnetic effect can be experienced. The magnetic field is represented by magnetic lines of force which form a closed loop. The greater the current through the conductor, the stronger is the magnetic field and vice versa. The magnetic field disappears as soon as the current is switched off or charges stop moving.

Consider a +*ve* charge $+q$ moving in a uniform magnetic field $\vec{B}$ with a velocity $\vec{V}$. Let the angle between $\vec{V}$ and $\vec{B}$ be θ as shown in Figure 1.10.

The magnitude that force $\vec{F}$ depends on is directly proportional to the magnitude of charge q, magnitude of the applied magnetic field and component of velocity acting perpendicular to the direction of magnetic field, i.e.,

$$F \propto qVB\sin\theta$$

$$\text{or } F = kqVB\sin\theta$$

where k is constant of proportionality, its value is found to be one.

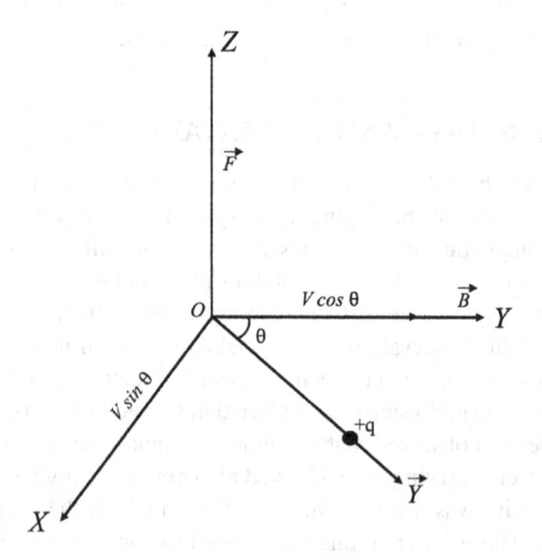

FIGURE 1.10 Diagram showing positive charge moving in a uniform magnetic field.

$$\therefore F = qVB\sin\theta \tag{1.33}$$

This equation can be written in the vector form as

$$\vec{F} = q\left(\vec{V} \times \vec{B}\right)$$

Furthermore, from the above equation, it is clear that if a charged particle is moving along or opposite to the direction of the magnetic field, it does not experience any force, and if the charge q moves at right angles to the magnetic field, then it experiences maximum force. Also, a stationary charged particle experiences no force in the magnetic field.

In 1820 Biot and Savart provided a quantitative relation between the magnetic field and its consequential current. When an electric current flows via a conductor, it generates a magnetic field around the conductor. The magnitude and direction of this field at a point due to this current can be obtained by a law, formulated on the basis of experiments by Biot and Savart. Consider a current element AB of a thin curved conductor XY through which a constant current I is maintained. Let dB be the magnitude of the magnetic field at P due to this current element of length dl as shown in Figure 1.11.

According to Biot-Savart law, the magnitude dB of magnetic field due to a current element at a point P is (1) directly proportional to the current I flowing through the conductor, (2) directly proportional to length dl of the current element, (3) directly proportional to the sin*e* of the angle between the direction of current element and vector $\vec{r}$, and (4) inversely proportional to the square of the distance of point P from the current element. It is also called inverse square law, i.e.,

$$dB \propto \frac{Idl\sin\theta}{r^2} \tag{1.34}$$

where θ is the angle between $\vec{r}$ and $\vec{dl}$, $\vec{r}$ is position vector of the observation point P with respect to the center O of the current element and r is distance of the observation point P from the midpoint O of the current element.

Equation (1.34) can be written,

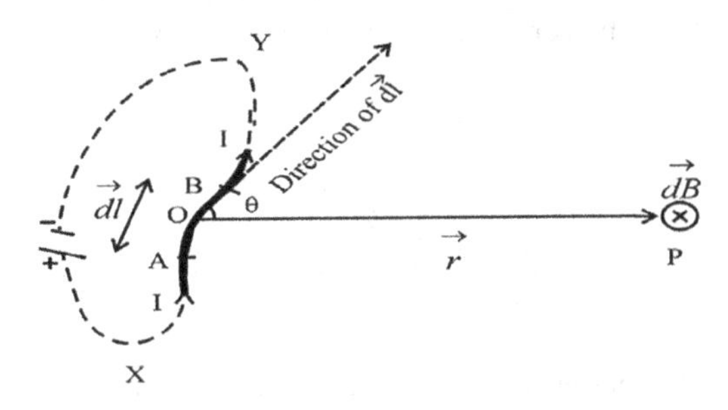

FIGURE 1.11 Magnetic field at a point P due to current element *dl*.

$$dB = K\frac{Idl\sin\theta}{r^2}$$

where K = constant of proportionality whose value depends on the medium between the observation point and the current element. For free space

$$K = \frac{\mu_0}{4\pi} = 10^{-7}TmA^{-1}$$

Where μ_0 is called the absolute permeability of free space.
Equation (1.34) can also be expressed as,

$$dB = \frac{\mu_0}{4\pi}\frac{Idl\sin\theta}{r^2} \tag{1.35}$$

When $\theta = 0$, then $\sin 0 = 0$, from the above equation the magnetic field is zero at all points on the axis of current element. Also, when $\theta = 90°$, then $\sin 90° = 1$, the magnetic field is maximum in a plane passing through the element and perpendicular to its axis.

In vector form, Equation (1.34) can be written as,

$$\overrightarrow{dB} = \frac{\mu_0}{4\pi}\frac{I\overrightarrow{dl}\sin\theta}{r^2}\hat{r}$$

Here $\hat{r}$ = unit vector in the direction of $\vec{r}$.

$$\text{Hence, } \vec{dB} = \frac{\mu_0}{4\pi}\frac{I\vec{dl}\sin\vec{\theta}}{r^2} \text{ or } \overrightarrow{dB} = \frac{\mu_0}{4\pi}\frac{I\left(\overrightarrow{dl}\times\vec{r}\right)}{r^3}$$

The direction of $\overrightarrow{dB}$ is the same as the direction of $\overrightarrow{dl}\times\vec{r}$ and is given by the right-hand rule of the cross-product of vectors.

1.11 EXPRESSION FOR DIVERGENCE OF MAGNETIC FIELD

In Figure 1.12, the current element $I\overrightarrow{dl}$ produces a magnetic field at a point $P(x, y, z)$ at a distance of 'r' from the current element and is given by

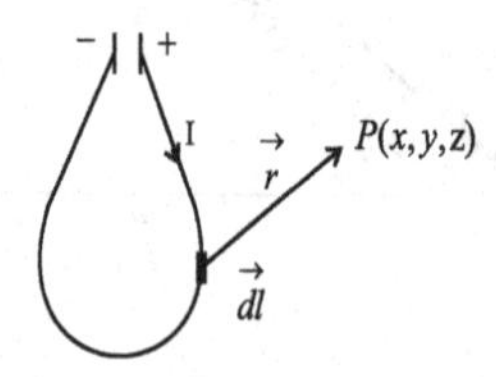

FIGURE 1.12 Current element dl producing a magnetic field at a point P.

$$\overrightarrow{dB} = \frac{\mu_0}{4\pi}\frac{I\overrightarrow{dl}\times\vec{r}}{r^3}\left(\text{Biot - Savart's law}\right)$$

The magnetic field at *P*, due to the whole current loop, is given by

$$\oint \vec{d}B = \oint \frac{\mu_0}{4\pi}\frac{I\left(\vec{d}l\times\vec{r}\right)}{r^3}$$

$$\text{or } \vec{B} = \frac{\mu_0}{4\pi}\oint\frac{I\left(\vec{d}l\times\vec{r}\right)}{r^3} \tag{1.36}$$

Taking divergence on both sides, we get

$$\vec{\nabla}\cdot\vec{B} = \frac{\mu_0}{4\pi}\oint\vec{\nabla}\cdot\frac{I\left(\vec{d}l\times\vec{r}\right)}{r^3}$$

$$\vec{\nabla}\cdot\vec{B} = \frac{\mu_0 I}{4\pi}\oint\vec{\nabla}\left(dl\times\frac{\vec{r}}{r^3}\right) \tag{1.37}$$

Using $\vec{\nabla}\cdot\left(\vec{A}\times\vec{B}\right) = \vec{B}\cdot\left(\vec{\nabla}\times\vec{A}\right) - \vec{A}\cdot\left(\vec{\nabla}\times\vec{B}\right)$, we get

$$\vec{\nabla}\cdot\vec{B} = \frac{\mu_0 I}{4\pi}\left[\oint\left(\frac{\vec{r}}{r^3}\right)\left(\vec{\nabla}\times\vec{d}l\right) - \vec{d}l\cdot\left(\vec{\nabla}\times\frac{\vec{r}}{r^3}\right)\right] \tag{1.38}$$

Since $\overrightarrow{dl}$ is not the function of *x*, *y* and *z*, therefore

$$\vec{\nabla}\times\vec{d}l = 0 \tag{1.39}$$

and $\vec{\nabla}\times\frac{\vec{r}}{r^3} = -\vec{\nabla}\times\vec{\nabla}\left(\frac{1}{r}\right)$

We know the curl of a gradient is zero

$$\therefore \vec{\nabla}\times\frac{\vec{r}}{r^3} = -\vec{\nabla}\times\vec{\nabla}\left(\frac{1}{r}\right) = 0 \tag{1.40}$$

Using Equations (1.38) and (1.39) in Equation (1.40), we get

$$\vec{\nabla}\cdot\vec{B} = 0 \tag{1.41}$$

Thus, divergence of $\vec{B}$ is zero.

Any vector whose divergence is zero is known as a solenoidal vector. Thus, a magnetic field vector $\vec{B}$ is a solenoid vector.

1.12 MAGNETIC SCALAR AND VECTOR POTENTIALS

1.12.1 Magnetic Scalar Potential

In the case of an electric field, we have

$$\vec{E} = -gradV = -\nabla V_E$$

where V_E is the electric potential, which is a scalar quantity, and $\vec{E}$ is the electric field intensity.

The electric field is a conservative field and hence

$$\vec{\nabla} \times \vec{E} = 0$$

$$\text{or } \vec{\nabla} \times -\vec{\nabla} V_E = 0$$

But curl $\vec{B} = \vec{\nabla} \times \vec{B} = 0$ only in the special case when the line integral $\oint \vec{B} \cdot \vec{dl} = 0$, i.e., when the line integral does not enclose a current. When the line integral encloses a current of current density $\vec{J}$, then

$$\vec{\nabla} \times \vec{B} = \mu_0 \vec{J}$$

In a current free space $\vec{J} = 0$ and therefore $\vec{\nabla} \times \vec{B} = 0$

Thus, for a current free space, we write

$$\vec{B} = -\vec{\nabla} V_m$$

Where, V_m is a scalar function called 'magnetic scalar potential'. $\vec{B}$ is the $-ve$ gradient of scalar potential only in current free space. $\vec{B}$ is not, in general, the $-ve$ gradient of a scalar potential. The condition for magnetic scalar potential to exist is that the current density vector $\vec{J} = 0$, i.e., it is a current free space.

1.12.2 Magnetic Vector Potential

A vector whose curl is equal to the magnetic induction $\vec{B}$ is known as a vector potential.

Let $\vec{A}$ be a vector whose curl is equal to $\vec{B}$

i.e.,
$$\vec{B} = \text{curl } \vec{A}$$

$$\vec{B} = \vec{\nabla} \times \vec{A}$$

where $\vec{A}$ is vector potential; hence, the divergence of a magnetic induction field $\left(\vec{\nabla} \cdot \vec{B}\right)$ is always zero, i.e.,

$$\vec{\nabla} \cdot \vec{B} = 0$$

if $\vec{A}$ is another vector, such that
$\vec{B} = \vec{\nabla} \times \vec{A}$, then
$\vec{\nabla} \cdot \vec{B} = \vec{\nabla} \cdot \vec{\nabla} \times \vec{A}$ is always equal to zero.

The vector $\vec{A}$ is called a magnetic vector potential. Hence, magnetic vector potential is defined as a vector function, the curl of which is equal to $\vec{B}$ the magnetic induction field.

1.13 EXPRESSION FOR MAGNETIC VECTOR POTENTIAL

According to Biot-Savart's law, the magnetic field due to current element $I\vec{dl}$ at position vector $\vec{r}$ is given by

$$\vec{dB} = \frac{\mu_0}{4\pi} I \frac{\vec{dl} \times r}{r^2} \tag{1.42}$$

Now in the current element $I\vec{dl}$, the direction of current is the same as that of $\vec{dl}$. Consider a volume distribution of current. For a volume element dV, we can write $dV = Sdl$, where S is the area of cross-section of the surface enclosing the volume

$$I\vec{dl} = \frac{I}{S} S\vec{dl} = \vec{J} Sdl \quad \because \frac{I}{S} = J \quad \text{and} \quad Sdl = dV$$

$$\text{or } I\vec{dl} = \vec{J}dV$$

because the direction of $\vec{J}$ is the same as that of $\vec{dl}$

∴ Equation (1.42) becomes

$$\vec{dB} = \frac{\mu_0}{4\pi} \frac{\vec{J}dV \times \hat{r}}{r^2} = \frac{\mu_0}{4\pi} \left(\frac{\vec{J} \times \hat{r}}{r^2} \right) dV$$

Therefore, the magnetic induction field $\vec{B}$ at any point at a distance r from the current element of whole volume dV is given by

$$\vec{B} = \frac{\mu_0}{4\pi} \iiint_V \left(\vec{J} \times \frac{\hat{r}}{r^2} \right) dV$$

$$\because \vec{B} = -\frac{\mu_0}{4\pi} \iiint_V \left(\vec{J} \times \vec{\nabla} \left[\frac{1}{r} \right] \right) dV \quad \because \frac{\hat{r}}{r^2} = -\vec{\nabla} \left(\frac{1}{r} \right)$$

$$\text{or } \because \vec{B} = -\frac{\mu_0}{4\pi} \iiint_V \left(\vec{J} \times \vec{\nabla} \left[\frac{1}{r} \right] \right) dV \quad \because \vec{A} \times \vec{B} = -\vec{B} \times \vec{A} = \frac{\mu_0}{4\pi} \iiint_V \left(\vec{\nabla} \left[\frac{1}{r} \right] \times \vec{J} \right) dV \tag{1.43}$$

Also we know $\vec{\nabla}\times\frac{\vec{J}}{r}=\left(\vec{\nabla}\times\vec{J}\right)\frac{1}{r}+\vec{\nabla}\left(\frac{1}{r}\right)\times\vec{J}$

For steady current $\vec{\nabla}\times\vec{J}=0$

$$\therefore \vec{\nabla}\times\frac{\vec{J}}{r}=\vec{\nabla}\left(\frac{1}{r}\right)\times\vec{J} \tag{1.44}$$

Using Equation (1.44) in Equation (1.43), we get

$$\vec{B}=\frac{\mu_0}{4\pi}\iiint_V\left(\vec{\nabla}\times\frac{\vec{J}}{r}\right)dV$$

$$\text{or } \vec{B}=\vec{\nabla}\times\frac{\mu_0}{4\pi}\iiint\frac{\vec{J}}{r}dV \tag{1.45}$$

Since $\vec{\nabla}\times\vec{B}\neq 0$ as in case of $\vec{E}$ and $\vec{\nabla}\cdot\vec{B}=0$. We define a vector potential $\vec{A}$ which is related to $\vec{B}$ by $\vec{B}=\vec{\nabla}\times\vec{A}$

$$\text{So } \vec{\nabla}\cdot\vec{B}=\vec{\nabla}\cdot\left(\vec{\nabla}\times\vec{A}\right)=0$$

$$\text{i.e., } \vec{B}=\text{curl } \vec{A}=\vec{\nabla}\times\vec{A}$$

From Equation (1.45)

$$\vec{\nabla}\times\vec{A}=\vec{\nabla}\times\frac{\mu_0}{4\pi}\iiint_V\left(\frac{\vec{J}}{r}\right)dV$$

$$\text{or } \vec{A}=\frac{\mu_0}{4\pi}\iiint_V\left(\frac{\vec{J}}{r}\right)dV$$

This is the required expression for magnetic vector potential.

1.14 MAGNETIZATION AND RELATION BETWEEN VARIOUS PARAMETERS

1.14.1 Magnetization

Every material is made up of atoms and each atom may be considered of electrons revolving about a positive nucleus. The electrons also rotate around their own axes. Thus, an internal magnetic field is formed by electrons revolving around the nucleus or spinning of electrons. Both electronic motions yield internal magnetic fields that are similar to the magnetic field produced by a current loop. The magnetic dipoles corresponding to each atom are randomly oriented in the magnetic substance in the absence of any external magnetic field. Thus, the net dipole moment of the magnetic substance is zero. However, if the same substance is placed in an external magnetic field, each dipole tends to align itself in the direction of the dipole moment. Such a magnetic substance is said to be magnetized. The process by which the substance is magnetized is known as 'magnetization'.

Ordinary conduction currents which arise because of the presence of batteries or other sources of electromagnetic field (e.m.f.) of electrical circuits that actually involve motion of electrons in a macroscopic path are called 'free currents'. Such currents can be started or stopped with a switch and measured with the help of an ammeter. On the other hand, the currents associated with molecular or atomic magnetic dipole moments are known as 'bound currents'. These arise due to orbital or spin motion of the electrons within the atom. Magnetization currents are also called 'bound currents'.

1.14.2 Relation between Various Parameters

For most paramagnetic and diamagnetic materials, the magnetization $\vec{M}$ is found proportional to the magnetizing field $\vec{H}$

$$\text{i.e., } \vec{M} \infty \vec{H} \text{ or } \vec{M} = \chi_m \vec{H} \tag{1.46}$$

Where χ_m is called 'magnetic susceptibility'. A material that obeys the relation $\vec{M} = \chi_m \vec{H}$ is called a 'linear medium'. We know that the magnetic field $\vec{B}$ is related to the magnetization $\vec{M}$ and magnetizing field $\vec{H}$ as

$$\vec{B} = \mu_0 \left(\vec{H} + \vec{M} \right) \tag{1.47}$$

Therefore, for a linear medium

$$\vec{B} = \mu_0 \left(\vec{H} + \chi_m \vec{H} \right) \quad \left| \text{Using Equations (1.46) and (1.47)} \right|$$

$$\text{or } \vec{B} = \mu_0 \vec{H} \left(1 + \chi_m \right) \tag{1.48}$$

Thus, we find that for a linear medium, the magnetic field $\vec{B}$ is found to be proportional to the magnetizing field $\vec{H}$

$$\vec{B} \infty \vec{H} \text{ or } \vec{B} = \mu \vec{H} \tag{1.49}$$

where μ is called the 'magnetic permeability' of the material.

Thus, the magnetic permeability of a medium is the ratio of the magnetic field (B) and the magnetizing field (H), i.e.,

$$\mu = \frac{B}{H} \tag{1.50}$$

Comparing Equations (1.47) and (1.48), we get

$$\mu_0 (1 + \chi_m) \vec{H} = \mu \vec{H}$$

$$\text{or } \mu = \mu_0 (1 + \chi_m) \tag{1.51}$$

This is the relation between the magnetic permeability (μ) and magnetic susceptibility χ_m of a linear medium. In a vacuum where there is no matter to magnetize, the magnetic susceptibility χ_m is zero and hence from Equation (1.51), we get

$$\mu = \mu_0 \tag{1.52}$$

This is why μ_0 is called the permeability of free space.
Thus, for vacuum (or air)

$$\mu_0 = \frac{B}{H} \text{ or } B = \mu_0 H \tag{1.53}$$

The ratio of the permeability of a medium to that of free space is called the 'relative permeability' (μ_r) of the medium, i.e.,

$$\mu_r = \frac{\mu}{\mu_0} \text{ or } \mu = \mu_0 \mu_r \tag{1.54}$$

∴ Equation (1.51) can be written as

$$\mu_0 \mu_r = \mu_0 (1 + \chi_m) \text{ or } \mu_r = 1 + \chi_m \tag{1.55}$$

$$\chi_m = \mu_r - 1 \tag{1.56}$$

This is the relation between the relative permeability (μ_r) and magnetic susceptibility χ_m of a linear medium.

1.15 CLASSIFICATION OF MAGNETIC MATERIALS

It has been well established that all materials are affected by external magnetic fields; some attain weak magnetic properties and some acquire strong magnetic properties. On the basis of their behavior in external magnetic fields, the various substances are classified into three categories as follows below.

1.15.1 Diamagnetic Substances

Diamagnetism happens in materials where the magnetic fields, due to electron motions of revolving and spinning, completely cancel each other. Thus, the permanent magnetic moment of every atom is zero, and the materials are weakly affected by an external magnetic field. For most diamagnetic materials (e.g., lead, bismuth, copper, diamond, silicon, sodium chloride), χ_m is of the order of 10^{-5}. In superconductors, perfect diamagnetism occurs: $\chi_m = 1$ or and $B = 0$ and $\mu_r = 0$ at absolute zero temperature. Thus superconductors cannot contain magnetic fields. Moreover, the diamagnetic substance loses its magnetism as soon as the external magnetic field is removed.

A diamagnetic substance is feebly repelled by a strong magnet; it is because a diamagnetic substance develops weak magnetization in the opposite direction of the applied magnetic field. Furthermore, when a diamagnetic substance is placed in a magnetic field, the magnetic lines of force prefer to pass through the surrounding air rather than through the substance. It is because the induced magnetic field in the diamagnetic substance opposes the external field. For this reason, the resultant field B inside the substance is less than the external field B_0. In addition, when a rod of diamagnetic substance is suspended freely in a uniform magnetic field, the rod comes to rest with its longest axis at right angles to the direction of the field. It happens because a diamagnetic substance is weakly repelled by a magnet. Also, when placed in a non-uniform magnetic field, a diamagnetic substance moves from stronger to weaker parts of the field.

In order to understand the cause of diamagnetism, we give the following description.

We know that each electron in an atom is revolving in an orbit around the nucleus. This revolving electron is equivalent to a tiny current loop. Therefore, each revolving electron has orbital magnetic dipole moment, $\vec{M}_l = \text{Current} \times \text{Area of the orbit}$. Further, an electron also spins about its own axis. This spinning motion produces an effective current loop and hence a spin magnetic moment $\vec{M}_S$. The vector sum of $\vec{M}_l$ and $\vec{M}_S$ provides the net magnetic dipole moment $\vec{M}$ to the atom.

In a diamagnetic substance, $\vec{M}_l$ and $\vec{M}_S$ cancel each other for every atom so that the net magnetic moment of the atom is zero. Therefore, motion of all electrons in the atom of a diamagnetic substance can be viewed as the motion of two electrons revolving with the same speed $\overrightarrow{V_0}$ in circular orbits of the same radius (r) but in the opposite directions. Because their magnetic moments are equal in magnitude and opposite in direction, the two magnetic moments cancel each other as shown in Figure 1.13.

Thus, in the absence of an external magnetic field, the atoms of a diamagnetic substance have no net magnetic moment. Hence, the substance does not exhibit diamagnetism. On the other hand, when an external uniform magnetic field $\vec{B}$ is applied perpendicular to the plane of the orbit and increasing into the paper, each electron experiences a magnetic force of magnitude $F_m = eVB$ which also contributes to the centripetal force. The direction of $\overrightarrow{F_m}$ can be determined by the right-hand rule for cross-product, for the electron moves in anticlockwise direction. The direction of $\overrightarrow{F_m}$

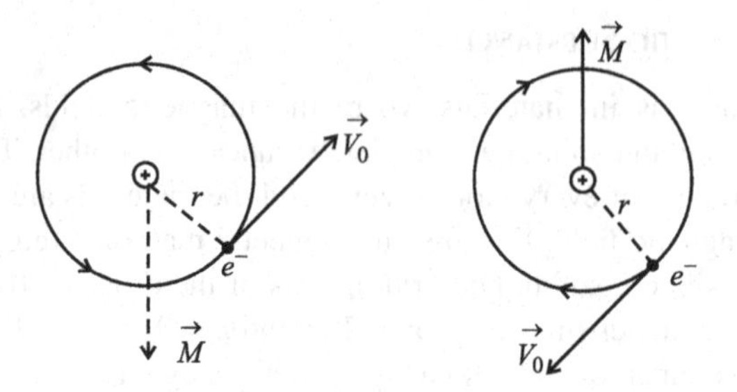

FIGURE 1.13 Two magnetic moments equal in magnitude and opposite in direction.

is radially outward, and for the electron moving clockwise, the direction of $\overrightarrow{F_m}$ is radially inward. Therefore, the speed of electron moving in anticlockwise direction decreases to $V_1\left(=\vec{V}_0-\Delta\vec{V}\right)$ and as a result, the magnetic moment of electron decreases to $\left(\vec{M}-\Delta\vec{M}\right)$. On the other hand, the speed of the electron moving in a clockwise direction increases to $\vec{V}_2\left(=\vec{V}_0+\Delta\vec{V}\right)$ and, therefore, the magnetic moment of electron increases to $(\vec{M}+\Delta\vec{M})$. The vector addition of these two magnetic moments gives rise to a net dipole magnetic moment $2\Delta\vec{M}$, directed opposite to the external magnetic field $\vec{B}$ as shown in Figure 1.14.

Thus, when a diamagnetic material is placed in an external magnetic field, an induced magnetic moment $\left(=2\Delta\vec{M}\right)$ is developed in the material which opposes the applied magnetic field. This accounts for the diamagnetic behavior of the material.

Note:

1. The relative permeability (μ_r) of a diamagnetic substance is always less than one (1).
2. The magnetic susceptibility (χ_m) of a diamagnetic substance has a small –*ve* value, e.g., –0.000015 for bismuth. It is because $\mu_r = 1 + \chi_m$ and $\mu_r < 1$
3. The magnetic susceptibility (χ_m) of a diamagnetic substance does not change with temperature.

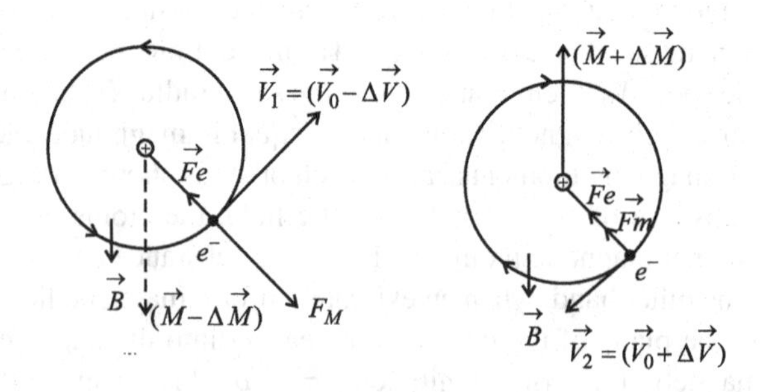

FIGURE 1.14 Vector analysis of magnetic moments of electrons in clockwise and anticlockwise directions.

1.15.2 Paramagnetic Substances

A paramagnetic material is weakly attracted by a strong magnet because a paramagnetic substance develops feeble magnetization in the direction of the applied magnetic field. When a paramagnetic substance is positioned in a magnetic field, the magnetic lines of force prefer to pass over the substance rather than through air. Therefore, the resultant field B inside the substance is more than the external field B_0. Also, when a rod of paramagnetic substance is suspended freely in a uniform magnetic field, the rod comes to rest with its longest axis along the direction of the external magnetic field. It happens because a paramagnetic substance is weakly attracted by a magnet. Furthermore, when placed in a non-uniform magnetic field, a paramagnetic substance moves from weaker to stronger parts of the field.

Paramagnetism is generally very weak, as only a very small fraction of the dipoles is aligned in the direction of the applied magnetic field. It is because the aligning process is counteracted by the tendency of the dipoles to be randomly oriented due to thermal motion. The fraction of the dipoles that line up with the field depends upon the strength of the field and the temperature. In fact, paramagnetism is quite sensitive to temperature. The lower the temperature, the stronger the paramagnetism and vice versa.

In order to understand the cause of paramagnetism, we give the following description.

In a paramagnetic substance, the individual atom/molecule/ion has a small net magnetic moment. In other words, the electron spins, and orbital motions have a net circulating current that is not zero. Therefore, the atom/molecule/ion as a whole has a net magnetic moment, i.e., each acts as a magnetic dipole.

In the absence of an external magnetic field, the dipoles of the paramagnetic substance are randomly oriented (Figure 1.15 (i)); therefore, the net magnetic moment of the substance is zero. Hence, the substance does not exhibit paramagnetism. On the other hand, when a paramagnetic substance is placed in an external magnetic field (Figure 1.15 (ii)), the dipoles are partially aligned in the direction of the applied field. Therefore, the substance is feebly magnetized in the direction of the applied magnetic field. This results in a weak attractive force on the substance.

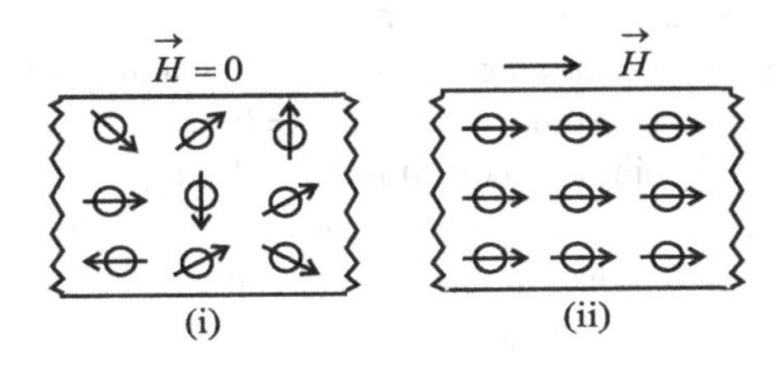

FIGURE 1.15 Diagram showing orientation of magnetic dipoles in the (i) absence and (ii) presence of an external magnetic field.

Note:

1. The relative permeability (μ_r) of a paramagnetic substance is always more than one (1).
2. The magnetic susceptibility (χ_m) of a paramagnetic substance has small *+ve* value, e.g., $+2.3 \times 10^{-5}$ for aluminum. It is because $\mu_r = 1 + \chi_m$ and $\mu_r > 1$.
3. The magnetic susceptibility (χ_m) of a paramagnetic substance varies inversely with the absolute temperature (T).

1.15.3 Ferromagnetic Substances

Ferromagnetism receives its name from word 'ferrous' meaning 'iron', which is considered to be the first metal to show attractive properties to when exposed to external magnetic fields. It is a unique magnetic behavior that is displayed by some materials like cobalt, iron, alloys, etc. It is a phenomenon by which these materials achieve permanent magnetism or they gain attractive powers. Ferromagnetism is a property that considers not only the chemical make-up of a material, but it also takes into account the microstructure and the crystalline structure.

Those substances which when placed in an external magnetic field are strongly magnetized in the direction of the applied external magnetic field are called 'ferromagnetic' substances, e.g., iron, nickel, cobalt, etc.

Since the strong induced magnetic field is in the direction of the applied magnetic field, the resultant magnetic field inside the ferromagnetic substance is very large; often thousands of times greater than the external field. It is clear that ferromagnetism is a very strong form of magnetism. When an external magnetic field is removed, some ferromagnetic substances retain magnetism. A ferromagnetic substance is strongly attracted by a magnet. Also, when a ferromagnetic substance is placed in a magnetic field, the magnetic field lines tend to coil into the substance. When a rod of ferromagnetic substance is suspended in a uniform magnetic field, it quickly aligns itself in the direction of the field. Furthermore, when placed in a non-uniform magnetic field, a ferromagnetic substance moves from weaker to stronger parts of the magnetic field.

In order to understand the cause of ferromagnetism, we give the following description.

Like paramagnetic substances, the atoms of ferromagnetic substances have a permanent magnetic moment. But in a ferromagnetic substance, the atoms do not act independently; rather they group magnetically into what are called 'domains'. Each domain contains a very large number of atoms ($=10^{17}$ to 10^{20} atoms) whose dipole moments are parallel and pointing in one direction. Therefore, each domain has a net magnetic dipole moment. The region of space over which the magnetic dipole moments of the atoms are aligned in the same direction is called a 'domain'.

In the absence of an external magnetic field, the domains of a ferromagnetic material are randomly oriented, as shown in Figure 1.16. In other words, within the domain, all magnetic moments are aligned in the same direction, but different domains are oriented randomly in different directions. The result is that one domain cancels the effect of the other, so that the net magnetic moment in the material is zero. Therefore, a ferromagnetic material does not exhibit magnetism in the normal state.

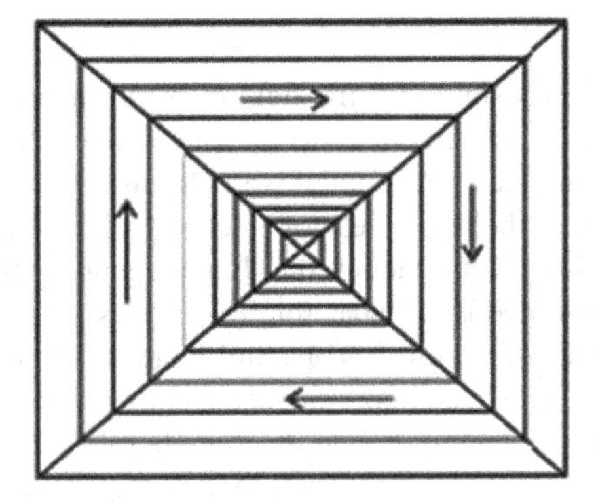

FIGURE 1.16 Diagram showing the ferromagnetic domains in absence of an external magnetic field.

On the other hand, when a ferromagnetic substance is placed in an external magnetic field, a net magnetic moment develops in the substance. By the displacement of boundaries of the domains, i.e., the domains that already happen to be aligned with the applied field may grow in size, whereas those oriented opposite to the external field reduce in size as shown in (Figure 1.17(i)) and by the rotation of the domains, i.e., the domains may rotate so that their magnetic moments are more or less aligned in the direction of the applied magnetic field, as shown in (Figure 1.17(ii)).

The result is that there is a net magnetic moment in the material in the direction of the applied field. Since the degree of alignment is very large even for a small external magnetic field, the magnetic field produced in the ferromagnetic material is often much greater than the external field.

Note:

1. The relative permeability (μ_r) of a ferromagnetic substance is very large, for example, the relative permeability of soft iron is about 8000.
2. The magnetic susceptibility (χ_m) of a ferromagnetic substance is +*ve*, having a very high value. It is because $\mu_r = 1 + \chi_m$ and $\mu_r \gg 1$ For this reason, ferromagnetic substances can be magnetized easily and strongly.

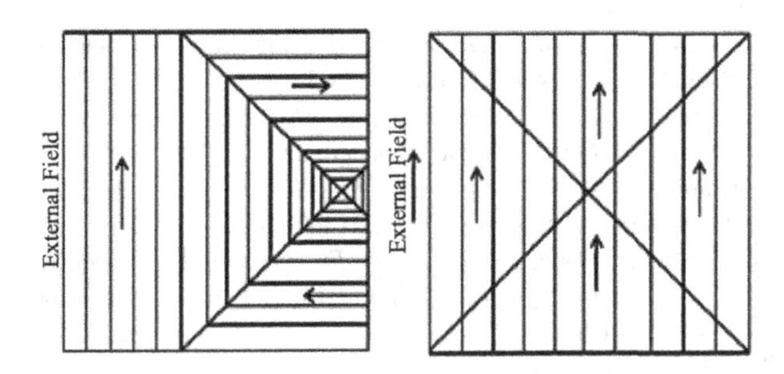

FIGURE 1.17 Ferromagnetic domains in presence of an external magnetic field.

1.16 SUMMARY

This chapter provides an insight into the field of electromagnetism. The basics of electrostatics, electric field, electric flux, Gauss's theorem of electrostatics in integral and differential form, and electric potential were discussed in depth in order to simplify the understanding of the basic ideas of electricity. The practical aspect of electrostatics is explained by introducing the fundamental topic like capacitance, capacitors, and energy stored by a capacitor. The expressions show that the energy can be regarded as either stored on the capacitor plates or stored in the electrostatic field between the plates. This is an important concept to grasp, as it shows the equivalence between a field theory approach and the more familiar 'circuits' approach. The question of how the current 'flowed' through an ideal capacitor is also presented. On the other hand, the other basic part of electromagnetism, i.e., 'magneto-statics' is introduced so that the readers are acquainted with fundamental concepts and principles of electromagnetism. The presentation of magnetic field and Biot-Savart law led to the important concept of magnetic scalar and vector potentials, in addition to the expressions for divergence of magnetic and magnetic vector potential. A separate section is presented for classification of magnetic materials, and each type is focused on properties, causes, and behavior under different conditions. A part of the chapter acquaints the reader with the practical aspects of magnetization. It can be concluded from the chapter that electric and magnetic fields are counterparts and provide a foundation for the rest of the book.

FURTHER READING

1. Dibner, Bern (2012). *Oersted and the Discovery of Electromagnetism.* Literary Licensing, LLC: Whitefish, MT. ISBN 978-1-258-33555-7.
2. Durney, Carl H., Johnson, Curtis C. (1969). *Introduction to Modern Electromagnetics.* McGraw-Hill: New York. ISBN 978-0-07-018388-9.
3. Feynman, Richard P. (1970). *The Feynman Lectures on Physics Vol. II.* Addison Wesley Longman: Boston, MA. ISBN 978-0-201-02115-8.
4. Grant, I.S., Phillips, W.R. (2008). *Electromagnetism* (2nd ed.). John Wiley & Sons: Hoboken, NJ. ISBN 978-0-471-92712-9.
5. Griffiths, David J. (1998). *Introduction to Electrodynamics* (3rd ed.). Prentice Hall International (UK) Ltd: London. ISBN 978-0-13-805326-0.
6. Jackson, John D. (1998). *Classical Electrodynamics* (3rd ed.). Wiley: Hoboken, NJ. ISBN 978-0-471-30932-1.
7. Moliton, André (2007). *Basic Electromagnetism and Materials.* Springer-Verlag New York, LLC: New York. ISBN 978-0-387-30284-3.
8. Purcell, Edward M. (1985). *Electricity and Magnetism Berkeley, Physics Course Volume 2* (2nd ed.). McGraw-Hill: New York. ISBN 978-0-07-004908-6.
9. Purcell, Edward M., Morin, David. (2013). *Electricity and Magnetism* (3rd ed.). Cambridge University Press: New York. ISBN 978-1-107-01402-2.
10. Rao, Nannapaneni N. (1994). *Elements of Engineering Electromagnetics* (4th ed.). Prentice Hall International (UK) Ltd.: London. ISBN 978-0-13-948746-0.
11. Rothwell, Edward J., Cloud, Michael J. (2001). *Electromagnetics.* CRC Press: Boca Raton, FL. ISBN 978-0-8493-1397-4.
12. Tipler, Paul (1998). *Physics for Scientists and Engineers: Vol. 2: Light, Electricity and Magnetism* (4th ed.). W.H. Freeman: New York. ISBN 978-1-57259-492-0.

2 Electromagnetic Theory

2.1 INTRODUCTION

The scalar wave theory was applied to the study of light before the development of the theory of electromagnetism. At that time, it was assumed that light waves were longitudinal, in analogy with sound waves; i.e., the wave displacements were in the direction of propagation. A further assumption, that light propagated through some type of medium, was made because the scientists of that time approached all problems from a mechanistic point of view. The scalar theory was successful in explaining diffraction, but problems arose in interpretation of the effects of polarization in interference experiments. Thomas Young (1773–1829) was able to resolve the difficulties by suggesting that light waves could be transverse, like the waves on a vibrating string. Using this idea, Augustin-Jean Fresnel (1788–1827) developed a mechanistic description of light that could explain the amount of reflected and transmitted light from the interface between two media.

Michael Faraday (1791–1867) observed in 1845 that a magnetic field would rotate the plane of polarization of light waves passing through the magnetized region. This observation led Faraday to associate light with electromagnetic radiation, but he was unable to quantify this association. Faraday attempted to develop electromagnetic theory by treating the field as lines pointing in the direction of the force that the field would exert on a test charge. The lines were given a mechanical interpretation with a tension along each line and a pressure normal to the line. James Clerk Maxwell (1831–1879) furnished a mathematical framework for Faraday. Maxwell identified light as "an electromagnetic disturbance in the form of waves propagated through the electromagnetic field according to electromagnetic laws" and demonstrated that the propagation velocity of light was given by the electromagnetic properties of the material.

Maxwell was not the first to recognize the connection between the electromagnetic properties of materials and the speed of light. Gustav Robert Kirchhoff (1824–1887) recognized in 1857 that the speed of light could be obtained from electromagnetic properties. Georg Friedrich Bernhard Riemann (1826–1866), in 1858, assumed that electromagnetic forces propagated at a finite velocity and derived a propagation velocity given by the electromagnetic properties of the medium. However, it was Maxwell who demonstrated that the electric and magnetic fields are waves that travel at the speed of light. It was not until 1887 that an experimental observation of electromagnetic waves, other than light, was obtained by Heinrich Rudolf Hertz. Furthermore, the electromagnetic fields are the consequence of rest and motion state of the electric charges. Electric field result is due to the positive and negative electric charges while magnetic field is due to the moving electric charges. Moreover, the electromagnetic field radiations are due to the coupling of time-varying electric and magnetic fields.

DOI: 10.1201/9781003213468-2

2.2 MAXWELL'S EQUATIONS

The basis of electromagnetic theory is Maxwell's equations. They allow the derivation of the properties of light. There are four fundamental equations illustrating the electric and magnetic fields and their interactions with charges and currents. These equations are named Maxwell's equations.

Gauss's law gives the relation between the charge and the produced electric field. The total electric flux coming out of a surface is equal to the charge enclosed by the closed surface divided by the permittivity and is named as Maxwell's first equation given as,

$$div \cdot \vec{E} = \vec{\nabla} \cdot \vec{E} = \frac{\rho}{\varepsilon_0} \tag{2.1}$$

The above equation is also known as Gauss's law of electrostatics.

Maxwell's second equation states that the divergence of the magnetic field is zero, or the magnetic monopoles does not exist, and it is given as,

$$div \cdot \vec{B} = \vec{\nabla} \cdot \vec{B} = 0 \tag{2.2}$$

The above equation is also known as Gauss's law of magnetostatics.

On the other hand, Faraday's law of electromagnetic induction is one of the fundamental laws of electromagnetism predicting the interaction between the magnetic field, electric current and the resultant production of electromotive force (e.m.f.), and is named as Maxwell's third equation given as,

$$\vec{\nabla} \times \vec{E} = -\frac{\partial \vec{B}}{\partial t} \tag{2.3}$$

Moreover, Maxwell-Ampere's circuital law states that a changing electric field creates a changing magnetic field and this changing magnetic field creates a changing electric field. Equations (2.4) and (2.5) below are referred to as Maxwell's fourth equation and is given as,

$$\vec{\nabla} \times \vec{B} = \mu_0 \vec{J} + \mu_0 \frac{\partial \vec{D}}{\partial t} \tag{2.4}$$

$$\vec{\nabla} \times \vec{H} = \vec{J} + \frac{\partial \vec{D}}{\partial t} \tag{2.5}$$

As we know, in free space, $\rho = 0$ and $\vec{J} = 0$. Therefore, Maxwell's equations are written as,

$$\vec{\nabla} \cdot \vec{E} = 0 \tag{2.6}$$

$$\vec{\nabla} \cdot \vec{B} = 0 \tag{2.7}$$

$$\vec{\nabla} \times \vec{E} = -\frac{\partial \vec{B}}{\partial t} \tag{2.8}$$

$$\vec{\nabla} \times \vec{B} = \mu_0 + \frac{\partial \vec{D}}{\partial t} \tag{2.9}$$

2.3 MAXWELL'S EQUATIONS IN MATTER

Let us first consider inside a dielectric medium. As per Gauss's law of electrostatics,

$$\vec{\nabla} \cdot \varepsilon_0 \vec{E} = \rho \tag{2.10}$$

Now, $\rho = \rho_b + \rho_f$

Where ρ is the total volume charge which is equal to the sum of volume charge density of bound charges (ρ_b) which arise due to the polarization of dielectric medium and volume charge density of free charges (ρ_f).

Since, $\rho_b = -\vec{\nabla} \cdot \vec{P}$

Here, $\vec{P}$ is the polarization vector.

Therefore,

$$\rho = -\vec{\nabla} \cdot \vec{P} + \rho_f \tag{2.11}$$

Using Equation (2.11) in (2.10), we get

$$\vec{\nabla} \cdot \varepsilon_0 \vec{E} = -\vec{\nabla} \cdot \vec{P} + \rho_f \tag{2.12}$$

Now we know that,

$$\vec{D} = \varepsilon_0 \vec{E} + \vec{P}$$

Here, $\vec{D}$ is the electric displacement vector.

Hence, Equation (2.12) becomes,

$$\vec{\nabla} \cdot \vec{D} = \rho_f \tag{2.13}$$

Now, let us consider inside a magnetic medium. As per Ampere-Maxwell's equation,

$$\vec{\nabla} \times \vec{B} = \mu_0 \vec{J} + \mu_0 \varepsilon_0 \frac{\partial \vec{E}}{\partial t} \tag{2.14}$$

Where,

$$\vec{J} = \vec{J}_m + \vec{J}_f \tag{2.15}$$

$\vec{J}_m = \vec{\nabla} \times \vec{M}$ is surface current density which results in the surface current, $\vec{M}$ represents the magnetization vector and $\vec{J}_m$ the current density.

Using Equation (2.15) in Equation (2.14), we get

$$\vec{\nabla} \times \vec{B} = \mu_0 \left(\vec{J}_m + \vec{J}_f \right) + \mu_0 \varepsilon_0 \frac{\partial \vec{E}}{\partial t}$$

$$\text{or } \vec{\nabla} \times \frac{\vec{B}}{\mu_0} - \vec{\nabla} \times \vec{M} = \vec{J}_f + \frac{\partial \left(\varepsilon_0 \vec{E} \right)}{\partial t}$$

$$\text{or } \vec{\nabla} \times \left(\frac{\vec{B}}{\mu_0} - \vec{M} \right) = \vec{J}_f + \frac{\partial \vec{D}}{\partial t} \tag{2.16}$$

$$\text{But, } \frac{\vec{B}}{\mu_0} - \vec{M} = \vec{H}$$

Therefore, Equation (2.16) becomes,

$$\vec{\nabla} \times \vec{H} = \vec{J}_f + \frac{\partial \vec{D}}{\partial t} \tag{2.17}$$

Hence, Maxwell's equations in matter are given as,

$$\vec{\nabla} \cdot \vec{D} = \rho_f \tag{2.18}$$

$$\vec{\nabla} \cdot \vec{B} = 0 \tag{2.19}$$

$$\vec{\nabla} \times \vec{E} = -\frac{\partial \vec{B}}{\partial t} \tag{2.20}$$

$$\vec{\nabla} \times \vec{H} = \vec{J}_f + \frac{\partial \vec{D}}{\partial t} \tag{2.21}$$

2.4 MAXWELL'S EQUATIONS IN INTEGRAL FORM

Maxwell's first equation in differential form is given by,

$$\vec{\nabla} \cdot \vec{D} = \rho$$

Integrating both sides over a volume V we get,

$$\iiint_V \vec{\nabla} \cdot \vec{D} dV = \iiint_V \rho dV \tag{2.22}$$

According to Gausss' divergence theorem,

$$\iiint_V \vec{\nabla} \cdot \vec{D} dV = \oiint \vec{D} \cdot \vec{dS}$$

Therefore, Equation (2.22) can be written as,

$$\oiint \vec{D} \cdot \vec{dS} = \iiint_V \rho dV \tag{2.23}$$

Since the volume integral of volume charge density(ρ) of free charges represents the total free charge (q) within the volume V, i.e.

$$\iiint_V \rho dV = q$$

Thus Equation (2.23) can be written as,

$$\oiint \vec{D} \cdot \vec{dS} = q$$

This equation is referred to as Maxwell's first equation in integral form. It states that the flux that belongs to electric displacement vector over a closed surface is equal to the total free charge enclosed within this surface.

Maxwells' second equation in differential form is given by,

$$\vec{\nabla} \cdot \vec{B} = 0$$

Integrating over a volume V, we get

$$\iiint_V \left(\vec{\nabla} \cdot \vec{B}\right) dV = 0 \tag{2.24}$$

According to Gauss's divergence theorem,

$$\iiint_V \left(\vec{\nabla} \cdot \vec{\mathrm{B}}\right) \mathrm{dV} = \oiint_S \vec{\mathrm{B}} \cdot \vec{\mathrm{dS}}$$

Therefore, Equation (2.24) becomes,

$$\oiint_S \vec{B} \cdot \vec{dS} = 0$$

This equation is referred to as Maxwell's second equation in integral form. It shows that the net magnetic flux through any closed surface is equal to zero.

Maxwell's third equation in differential form is given by,

$$\vec{\nabla} \times \vec{E} = \frac{-\partial \vec{B}}{\partial t}$$

Integrating both sides over a surface S bounded by a curve, we get

$$\oiint_S \left(\vec{\nabla} \times \vec{E}\right) \cdot \vec{ds} = -\oiint_S \left(\frac{\partial \vec{B}}{\partial t}\right) \cdot \vec{ds} \tag{2.25}$$

But, according to the Stokes's theorem,

$$\oiint_S \left(\vec{\nabla} \times \vec{E}\right) \cdot \vec{ds} = \oint_C \vec{E} \cdot \vec{dl}$$

Therefore, Equation (2.25) can be written as,

$$\oint_C \vec{E} \cdot \vec{dl} = -\frac{\partial}{\partial t} \oiint_S \vec{B} \vec{ds}$$

This equation is Maxwell's third equation in integral form. It shows that line integral of electric field strength E over a closed loop is equal to $-ve$ of time rate of change of magnetic flux associated with the path.

Maxwell's fourth equation in differential form is given by,

$$\vec{\nabla} \times \vec{H} = \vec{J} + \frac{\partial \vec{D}}{\partial t}$$

Integrating both sides over a surface S bounded by a closed loop, we get

$$\oiint_S \left(\vec{\nabla} \times \vec{H}\right) \cdot \vec{ds} = \oiint_S \left(J + \frac{\partial \vec{D}}{\partial t}\right) \cdot \vec{ds} \tag{2.26}$$

According to Stokes's theorem,

$$\oiint_S \left(\vec{\nabla} \times \vec{H}\right) \cdot \vec{ds} = \oint_C \vec{H} \cdot \vec{dl}$$

Therefore, Equation (2.26) can be written as,

$$\oint_C \vec{H} \cdot \vec{dl} = \oiint_S \left(J + \frac{\partial \vec{D}}{\partial t} \right) \cdot \vec{dS}$$

This equation is referred to as Maxwell's fourth equation in integral form. It shows that the magnetomotive force (m.m.f.) around a closed path (C) is equal to the sum of flux of current density $\left(\vec{J}\right)$ and displacement current density $\frac{\partial \vec{D}}{\partial t}$ over a surface (S) bounding the closed loop (C).

2.5 DERIVATION OF MAXWELL'S EQUATIONS

Let us consider a volume V enclosed by the surface (S) containing a dielectric medium. The total charge in a dielectric medium consists of free charge plus polarization charge. If (ρ) and (ρ_P) are the free charge density and polarization charge density at a point in a small volume element (dV), then Gauss's law can be expressed as,

$$\oiint_S \vec{E} \cdot \vec{dS} = \frac{1}{\varepsilon_0} \iiint_V (\rho + \rho_P) dV \tag{2.27}$$

But polarization charge density, $\rho_P = -\vec{\nabla} \cdot \vec{P}$.

Therefore, Equation (2.27) can be written as,

$$\oiint_S \vec{E} \cdot \vec{dS} = \frac{1}{\varepsilon_0} \iiint_V \left(\rho - \vec{\nabla} \cdot \vec{P}\right) dV \tag{2.28}$$

According to Gauss's divergence theorem,

$$\oiint_S \vec{E} \cdot \vec{dS} = \iiint_V \left(\vec{\nabla} \cdot \vec{E}\right) dV \tag{2.29}$$

From Equations (2.28) and (2.29), we get

$$\varepsilon_0 \iiint_V \left(\vec{\nabla} \cdot \vec{E}\right) dV = \iiint_V \rho dV - \iiint_V \vec{\nabla} \cdot \vec{P}) dV$$

$$\text{or} \quad \iiint_V \left(\vec{\nabla} \cdot \varepsilon_0 \vec{E}\right) dV = \iiint_V \left(\vec{\nabla} \cdot \vec{P}\right) dV = \iiint_V \rho dV$$

$$\text{or} \quad \iiint_V (\vec{\nabla} \cdot (\varepsilon_0 \vec{E} + \vec{P}) dV = \iiint_V \rho dV$$

But, $\varepsilon_0 \vec{E} + \vec{P} = \vec{D}$ = electric displacement vector.

Therefore, $\iiint_V (\vec{\nabla} \cdot \vec{E}) dV = \iiint_V \rho dV$

$$\text{or} \quad \iiint_V (\vec{\nabla} \cdot \vec{D} - \rho) dV = 0$$

This equation holds for any arbitrary volume V. It can further be written as,

$$\vec{\nabla} \cdot \vec{D} = \rho$$

We know that due to the non-existence of isolated magnetic poles and magnetic currents, the magnetic lines of force are either off the infinity or exist in closed curves. The number of magnetic lines of force entering or leaving any arbitrary closed surface is exactly the same. It means that the flux of magnetic induction across any closed surface is always zero, i.e.,

$$\oiint_S \vec{B} \cdot \vec{dS} = 0 \tag{2.30}$$

According to Gauss's divergence theorem,

$$\oiint_S \vec{B} + \vec{dS} = \iiint_V (\vec{\nabla} \cdot \vec{B}) dV$$

Therefore, Equation (2.30) can be written as,

$$\iiint_V (\vec{\nabla} \cdot \vec{B}) dV$$

This equation is true for any arbitrary volume V and can hold good if its integral is zero. Therefore, $\vec{\nabla} \cdot \vec{B} = 0$

On the other hand, according to Faraday's law of electromagnetic induction, the induced e.m.f. (e) is always equal to $-ve$ of time rate of change of magnetic flux ϕ, i.e.,

$$e = -\frac{d\phi}{dt} \tag{2.31}$$

But magnetic flux ϕ linked with a surface (S) is given by,

$$\phi = \iint_S \vec{B} \cdot \vec{dS}$$

Therefore, Equation (2.31) can be written as,

$$e = -\frac{d}{dt}\iint_S \vec{B}\cdot\vec{dS}$$

Since surface is fixed in space, hence only $\vec{B}$ changes with time,
Therefore,

$$e = \iint_S \frac{\partial \vec{B}}{\partial t}\cdot \vec{d}S \tag{2.32}$$

But e.m.f. is given by the work done in moving a unit charge around the closed loop C. Thus, if $\vec{E}$ is the electric field intensity at a small element dl of loop, then we have

$$e = \oint_C \vec{E}\cdot\vec{dl} \tag{2.33}$$

Comparing Equations (2.32) and (2.33) we get

$$\oint_C \vec{E}\cdot\vec{dl} = -\iint_S \frac{\partial \vec{B}}{\partial t}\cdot\vec{dS} \tag{2.34}$$

According to Stokes's theorem

$$\oint_C \vec{E}\cdot\vec{dl} = -\iint_S \left(\vec{\nabla}\times\vec{E}\right)\cdot\vec{dS}$$

Therefore, Equation (2.34) can be written as,

$$\iint_S \left(\vec{\nabla}\times\vec{E}\right)\cdot\vec{dS} = -\iint_S \frac{\partial \vec{B}}{\partial t}\cdot\vec{dS}$$

$$\text{or} \quad \iint_S \left(\vec{\nabla}\times\vec{E}\right)\cdot\vec{dS} = +\iint_S \frac{\partial \vec{B}}{\partial t}\cdot\vec{dS} = 0$$

$$\text{or} \quad \iint_S \left(\vec{\nabla}\times\vec{E} + \frac{\partial \vec{B}}{\partial t}\right)\cdot\vec{dS} = 0$$

This equation holds true for any arbitrary surface, S, and can hold good if its integral is zero.

Therefore, $\vec{\nabla}\times\vec{E} = -\dfrac{\partial \vec{B}}{\partial t}$

Moreover, as per Ampere's circuital law

$$\oint_C \vec{B} \cdot \vec{dl} = \mu_0 I$$

For vacuum, $\vec{B} = \mu_0 \vec{H}$

Therefore, $\oint_C \vec{H} \cdot \vec{dl} = I$

As Maxwell pointed out, if the current I in the above equation is taken as a simple conduction current, then it provides inconsistent results in problems involving time variations of electric field. He introduced a new current named displacement current I_D on the right-hand side (R.H.S.) of this equation to establish the consistency of results.

The improved form of Ampere's circuital law thus becomes,

$$\oint_C \vec{H} \cdot \vec{dl} = I + I_D \tag{2.35}$$

According to Stokes's theorem

$$\oint_C \vec{H} \cdot \vec{dl} = \iint_S \left(\vec{\nabla} \times \vec{H}\right) \cdot \vec{dS}$$

Also, $I + I_D = \iint_S \left(\vec{J} + \vec{J_D}\right) \cdot \vec{dS}$

Therefore, Equation (2.35) becomes,

$$\iint_S \left(\vec{\nabla} \times \vec{H}\right) \cdot \vec{dS} = \iint_S \left(\vec{J} + \vec{J_D}\right) \cdot \vec{dS}$$

$$\text{or} \quad \iint_S \left(\vec{\nabla} \times \vec{H}\right) \cdot - \iint_S \left(\vec{J} + \vec{J_D}\right) \cdot \vec{dS} = 0$$

This equation holds good for any arbitrary surface (S); therefore, its integral must be zero.

Therefore, $\left(\vec{\nabla} \times H\right) - \left(\vec{J} + \vec{J_D}\right) = 0$ or

$$\vec{\nabla} \times \vec{H} = \vec{J} + \vec{J_D} \tag{2.36}$$

Taking divergence on both sides, we get

$$div\left(\vec{\nabla} \times \vec{H}\right) = div\left(\vec{J} + \vec{J_D}\right) \tag{2.37}$$

We know that divergence of a curl is always zero.

Therefore, Equation (2.37) can be written as,

$$\vec{\nabla}\cdot\vec{J}+\vec{\nabla}\cdot\vec{J}_D=0 \tag{2.38}$$

Now equation of continuity is given by

$$\vec{\nabla}\cdot\vec{J}+\frac{\partial\rho}{\partial t}=0$$

$$\text{or } \vec{\nabla}\cdot\vec{J}=-\frac{\partial\rho}{\partial t}$$

Therefore, Equation (2.38) can be written as,

$$-\frac{\partial\rho}{\partial t}+\vec{\nabla}\cdot\overrightarrow{J_D}=0$$

$$\text{or } \vec{\nabla}\cdot\overrightarrow{J_D}=\frac{\partial\rho}{\partial t} \tag{2.39}$$

Also, Maxwell's first equation is given by

$$\vec{\nabla}\cdot\vec{\mathrm{D}}=\rho$$

Therefore, Equation (2.39) can be written as,

$$\vec{\nabla}\cdot\overrightarrow{J_D}=\frac{\partial}{\partial t}\left(\vec{\nabla}\cdot\vec{D}\right)$$

$$\text{or } \vec{\nabla}\cdot\overrightarrow{J_D}=\vec{\nabla}\cdot\frac{\partial\vec{D}}{\partial t}$$

$$\text{or } \overrightarrow{J_D}=\frac{\partial\vec{D}}{\partial t} \tag{2.40}$$

Using Equation (2.40) in Equation (2.36), we get

$$\vec{\nabla}\times\vec{H}=\vec{J}+\frac{\partial\vec{D}}{\partial t}$$

which is Maxwell's fourth equation of electromagnetic theory.

2.6 ELECTROMAGNETIC WAVES

Faraday's law of electromagnetic induction states that a time-varying magnetic field produces a time varying electric field. It means that when either of the magnetic field or electric field varies with time, then the other field is induced in the space which changes itself with time. Thus, a time-varying magnetic field produces a time-varying electric field. That changing electric field will in turn produce a changing magnetic field, and so on. The interaction between the time-varying magnetic and electric fields results in the generation of electromagnetic waves propagating in the space whose direction of propagation is perpendicular to both the electric and magnetic fields. Furthermore, Maxwell proved that the planes belonging to the variations in electric and magnetic are mutually perpendicular to each other. These waves are called 'electromagnetic waves' and do not require material medium for the propagation.

The electromagnetic waves consist of sinusoidal variation of electric and magnetic fields at right angles to each other as well as at right angles to the direction of propagation waves. Figure 2.1 shows an electromagnetic wave propagating along the X-axis. Here, electric field vector $\left(\vec{E}\right)$ and magnetic field vector $\left(\vec{B}\right)$ are vibrating along the Y-axis and Z-axis, respectively, while the wave travels along the X-axis.

The following points are worth noting about electromagnetic waves:

A. The electric and magnetic fields are always perpendicular to each other and both are perpendicular to the direction of propagation of wave. Therefore, electromagnetic waves are transverse in nature, and examples of electromagnetic waves are radio waves, infrared waves, x-rays, light, etc.

B. Maxwell found the speed of electromagnetic waves in free space as,

$$C = \frac{1}{\sqrt{\mu_0 \varepsilon_0}}$$

Where, μ_0 and ε_0 are, respectively, the permeability and permittivity of the free space.

Now, $\mu_0 = 4\pi \times 10^{-7} \frac{N}{A^2}, \varepsilon_0 = 8.85 \times 10^{-12}\ m^{-3}\ \text{kg}^{-1}\ S^4\ A^2$

Therefore, $C = \frac{1}{\sqrt{\left(4\pi \times 10^{-7}\right)\left(8.85 \times 10^{-12}\right)}} = 3 \times 10^8\ m/\text{sec}$

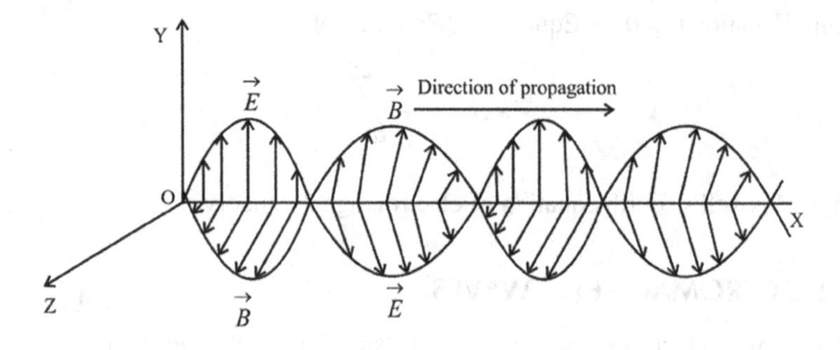

FIGURE 2.1 Electromagnetic wave propagating along the X-axis.

This is a remarkable result, as it is precisely equal to the measured speed of light in free space. Hence, light is an electromagnetic wave.

C. The velocity of electromagnetic waves in a medium is given by

$$v = \frac{1}{\sqrt{\mu\varepsilon}}$$

Where, μ = Absolute permeability of the medium
ε = Absolute permittivity of the medium.

Now, $\mu = \mu_0\mu_r$ and $\varepsilon = \varepsilon_0\varepsilon_r$. Here, μ_r and ε_r are the relative permeability and relative permittivity of the medium, respectively. Therefore,

$$v = \frac{1}{\sqrt{\mu_0\mu_r\varepsilon_0\varepsilon_r}} = \frac{1}{\sqrt{\mu_0\varepsilon_0}} \times \frac{1}{\sqrt{\mu_r\varepsilon_r}} = \frac{C}{\sqrt{\mu_r\varepsilon_r}}$$

2.7 EQUATION OF CONTINUITY

In electromagnetic theory, the continuity equation is an empirical law expressing conservation of charge. The equation of continuity is the relation between current density vector $\vec{J}$ and volume charge density ρ and this equation holds at each point of space and time.

Let us consider a small area element $\vec{dS}$ of a closed surface S, as shown in Figure 2.2. If P is a point on the area element $\vec{dS}$ and $\vec{J}$ the current density vector at P, then the charge flowing out of the area $\vec{dS}$ per second is given as,

$$dI = \vec{J} \cdot \vec{dS} \tag{2.41}$$

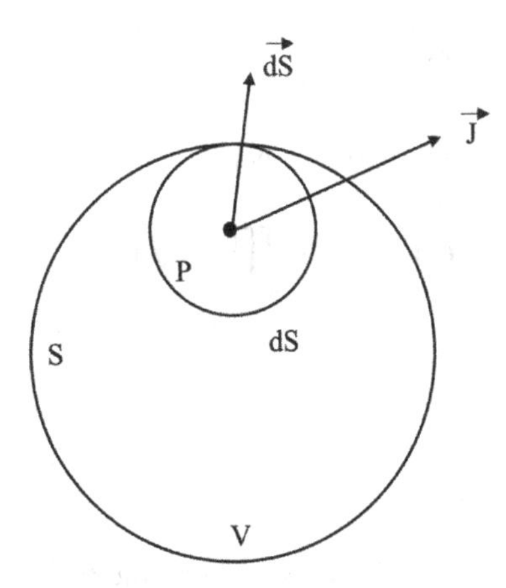

FIGURE 2.2 Small area element $\vec{dS}$ of a closed surface S.

Therefore, total charge crossing the closed surface S outward per unit time is given by

$$I = \oiint_S \vec{J} \cdot \vec{ds} \tag{2.42}$$

If ρ is the volume density of charge at any point, the total charge within the volume V enclosed by the surface S can be written as,

$$q = \iiint_V \rho dV \tag{2.43}$$

Since the current is flowing outward and the charge within the enclosed surface is decreasing with time, then by using Equation (2.43), the time rate of decrease of charge is given as,

$$\frac{-dq}{dt} = I = -\frac{\partial}{\partial t} \iiint_V \rho dV \tag{2.44}$$

According to the principle of conservation of charge, the total charge crossing the closed surface S per second is equal to the rate of decrease of charge in the enclosed volume, i.e.,

$$\oiint_S \vec{J} \cdot \vec{ds} = \iiint_V \frac{\partial \rho}{\partial t} dV \tag{2.45}$$

But according to Gauss's divergence theorem,

$$\oiint_S \vec{J} \cdot \vec{ds} = \iiint_V \left(\vec{\nabla} \cdot \vec{J} \right) dV \tag{2.46}$$

From Equations (2.45) and (2.46), we get

$$-\iiint_V \frac{\partial \rho}{\partial t} dV = \iiint_V \left(\vec{\nabla} \cdot \vec{J} \right) dV$$

$$\text{or} \quad \frac{\partial \rho}{\partial t} = \vec{\nabla} \cdot \vec{J}$$

Therefore, $\vec{\nabla} \cdot \vec{J} + \frac{\partial \rho}{\partial t} = 0$

This equation is referred as the equation of continuity which denotes the physical fact of the charge conservation. The term $\vec{\nabla} \cdot \vec{J}$ denotes the limiting value of net outward flow of electric current per unit area, while the term $\frac{\partial \rho}{\partial t}$ denotes the rate of change of charges per unit volume.

2.8 DISPLACEMENT CURRENT

The quantity $\partial D/\partial t$ that appears in Maxwell's equations in electromagnetism, referred to as displacement current density, is defined in terms of the rate of change of electric displacement field 'D'. The unit of displacement current density is the same as that of electric current density, and it is a source of the magnetic field like the actual current. However, it is a time-varying electric field, not the moving electric charge. In other words, it is the current through the gap between the plates of the capacitor due to the changing electric field and is denoted by I_d. Let us consider a parallel plate capacitor, AB, and a resistance, R, as shown in Figure 2.3(i).

Now, if the plate, A, of the capacitor carries a positive charge and the plate, B, an equal negative charge, then capacitor will discharge itself through the resistance, R, and a varying current will flow through the circuit when the key, K, is closed.

Let I be the electric current at any instant of time during the discharging of the capacitor and $\vec{B}$ the magnetic field at a point on the surface $\vec{S_1}$ sufficiently far away from the capacitor plates. If current I flows through the surface S_1 and $\vec{J}$ is the corresponding current density, then,

$$\text{curl } \vec{B} = \vec{\nabla} \times \vec{B} = \mu_0 \vec{J} \text{ (Ampere's circuital law in differential form).}$$

Taking surface integral on both sides, we get

$$\iint_{S_1} Curl\ \vec{B} \cdot \vec{d}S = \mu_0 \iint_{S_1} \vec{J} \cdot \vec{d}S \tag{2.47}$$

According to Stokes's theorem in vector analysis, if C is the linear boundary enclosing the surface.

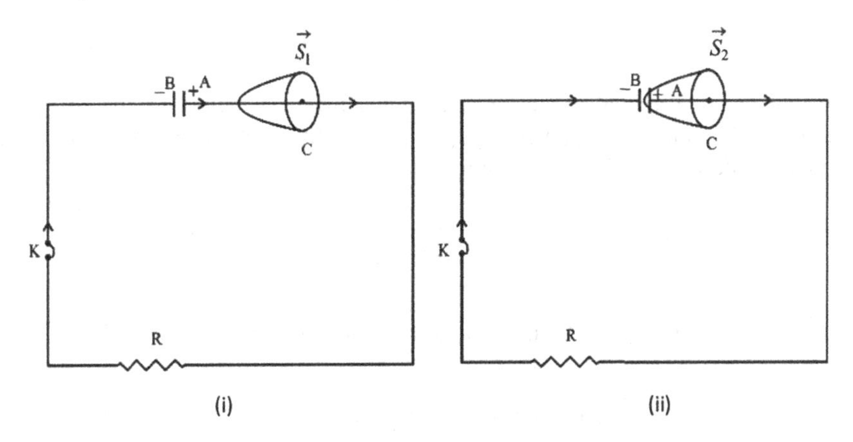

FIGURE 2.3 Parallel plate capacitor (i) outside and (ii) inside the linear boundary C.

Then, $\oint_C \vec{B}\cdot\vec{dl} = \iint_{S_1} Curl\ \vec{B}\cdot\vec{dS}$

Therefore, from Equation (2.47), we have

$$\oint_C \vec{B}\cdot\vec{dl} = \mu_0 \iint_{S_1} \vec{J}\cdot\vec{dS} = \mu_0 I$$

Now, consider a surface $\vec{S}_2$ as shown in Figure 2.3(ii). No current is flowing through the surface $\vec{S}_2$, but it is enclosed by the same linear boundary C. Therefore, for surface S_2

$$\oint_C \vec{B}\cdot\vec{dl} = \mu_0 \iint_{S_1} \vec{J}\cdot\vec{dS} = \mu_0 I = 0$$

The surfaces S_1 and S_2 are both bounded by same linear boundary C, and whereas $\oint_C \vec{B}\cdot\vec{dl}$ for the surface S_1 has a value $\mu_0 I$ and $\oint_C \vec{B}\cdot\vec{dl}$ for the surface S_2=0. This is not possible.

In other words, something is missing in the equation,

$$\text{curl } \vec{B} = \vec{\nabla}\times\vec{B} = \mu_0\vec{J}$$

and it should be in fact of the form,

$$\text{curl } \vec{B} = \vec{\nabla}\times\vec{B} = \mu_0\vec{J} + (\text{additional term})$$

Now, according to Faraday's law of electromagnetic induction in the differential form,

$$\vec{\nabla}\cdot\vec{E} = -\frac{\partial\vec{B}}{\partial t}$$

It indicates the mutual existence of changing magnetic field with the changing electric field. Symmetrically, a changing electric field should be accompanied by a magnetic field. Now we have a relation,

$$\vec{\nabla}\cdot\vec{B} = \mu_0\varepsilon_0\frac{\partial\vec{E}}{\partial t}$$

The above relation provides us the missing term, i.e.,

$$\text{curl } \vec{B} = \vec{\nabla}\times\vec{B} = \mu_0\left[\vec{J} + \varepsilon_0\frac{\partial\vec{E}}{\partial t}\right].$$

2.8.1 Expression for Displacement Current

The plates of a parallel plate capacitor are connected by connecting wires to a battery as shown in Figure 2.4. The charge starts to rise on the plates of the capacitor, which results in the increase of electric field between the plates of the capacitor with time.

Let q be the instantaneous charge on the positive plate of the capacitor, which is given by,

$$q = CV \tag{2.48}$$

Where,

$$C = \frac{\varepsilon_0 A}{d} \tag{2.49}$$

Where A is the area of each plate of the capacitor, and d is the distance of separation between the plates of the capacitor.

If V is the potential difference between the plates of the capacitor, then the magnitude of the electric field in the gap between the plates of the capacitor is given by,

$$E = \frac{\mathrm{V}}{d}$$

$$\text{or } \mathrm{V} = Ed \tag{2.50}$$

Using Equations (2.49) and (2.50) in Equation (2.48), we get

$$q = \left(\frac{\varepsilon_0 A}{d}\right) Ed$$

$$\text{or } q - \varepsilon_0 AE$$

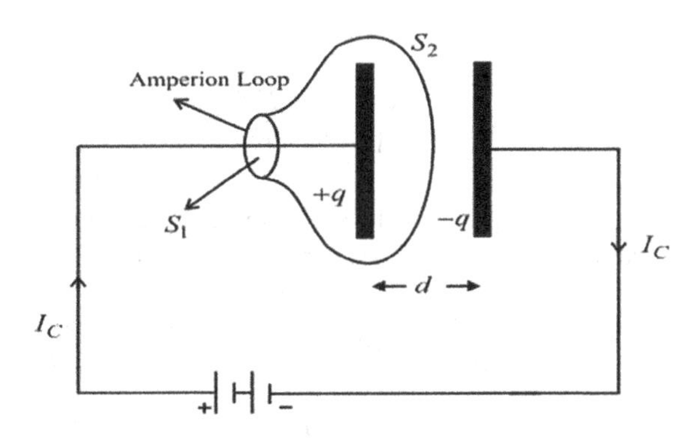

FIGURE 2.4 Parallel plate capacitor connected by connecting wires to a battery.

$$\text{or } E = \frac{q}{\varepsilon_0 A} \tag{2.51}$$

The electric flux through the surface S_2 of loop is given by,

$$\phi_E = EA \tag{2.52}$$

Substituting Equation (2.51) in Equation (2.52), we get

$$\phi_E = \left(\frac{q}{\varepsilon_0 A}\right) A$$

$$\text{or } \phi_E = \frac{q}{\varepsilon_0} \tag{2.53}$$

Therefore, $\dfrac{d\phi_E}{dt} = \dfrac{1}{\varepsilon_0}\dfrac{dq}{dt}$

$$\text{or } I_d = \varepsilon_0 \frac{d\phi_E}{dt}$$

Using Equation (2.53), we get

$$I_d = \varepsilon_0 \frac{d}{dt}(EA)$$

$$\text{or } I_d = \varepsilon_0 A \frac{dE}{dt} \tag{2.54}$$

which is the expression for displacement current.

Now, displacement current density is given as,

$$J_d = \frac{I_d}{A}$$

$$\text{or } J_d = \frac{I_d}{A} \times \varepsilon_0 A \frac{\partial E}{\partial t} \tag{2.55}$$

If a dielectric medium of permittivity ε fills the gap between the plates of the capacitor, then Equations (2.54) and (2.55) can be written as

$$I_d = \varepsilon A \frac{dE}{dt} \tag{2.56}$$

$$\text{and } J_d = \varepsilon \frac{dE}{dt} \tag{2.57}$$

Thus, Maxwell Ampere's circuital law can be written as

$$\oint \vec{B} \cdot \vec{dl} = \mu_0 \left[I_c + I_d \right] \tag{2.58}$$

Substituting Equation (2.56) in Equation (2.58), we get

$$\oint \vec{B} \cdot \vec{dl} = \mu_0 \left[I_c + \varepsilon_0 A \frac{dE}{dt} \right]$$

2.9 POYNTING VECTOR AND POYNTING THEOREM

The Poynting vector denotes the energy transfer per unit area per unit time associated with the electromagnetic field. When the electromagnetic wave travels from one point to another point in space, it transports energy from one point to another point. The rate of energy transfer can be expressed in terms of electric and magnetic fields constituting electromagnetic wave by a vector $\vec{S}$ called 'Poynting vector'.

The Pointing vector is defined as the cross-product of electric field vector $\vec{E}$ and magnetic field vector $\vec{H}$. It is denoted by $\vec{S}$. It is named after J. H. Poynting.

$$\vec{S} = \frac{1}{\mu_0} \left(\vec{E} \times \vec{B} \right) = \left(\vec{E} \times \vec{H} \right) \tag{2.59}$$

The direction of a Poynting vector is perpendicular to the plane containing $\vec{E}$ and $\vec{B}$. The direction of a Poynting vector $\vec{S}$ is the direction in which energy is transferred by electromagnetic wave. Now, let electric field $\vec{E}$ vary along the Y-axis and magnetic field $\vec{B}$ varies along the Z-axis as shown in Figure 2.5; therefore, Equation (2.59) can be written as,

$$\vec{S} = E\hat{j} \times H\hat{k}$$

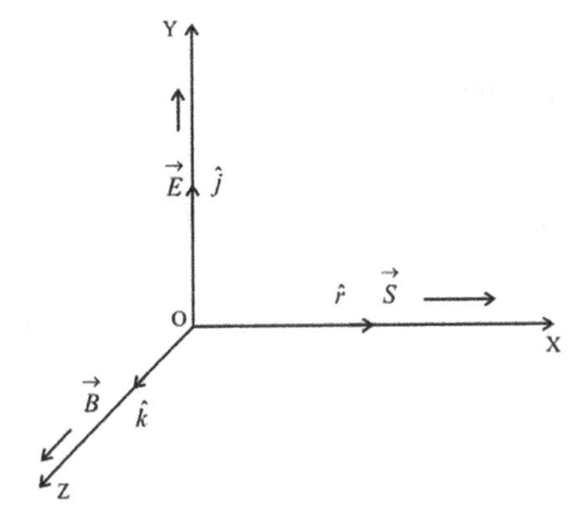

FIGURE 2.5 Varying electric and magnetic fields along the Y- and X-axis.

$$\text{or } \vec{S} = EH\left(\hat{j} \times \hat{k}\right)$$

$$\text{or } \vec{S} = (EH)\hat{i}$$

Therefore, the direction of Poynting vector is along the X-axis. It means that it is perpendicular to the plane containing $\vec{E}$ and $\vec{B}$ or $\vec{H}$.

The S.I. unit of a Poynting vector is Wm^{-2} or J/S/m^2. The Poynting vector gives the electrical power per unit area. Hence, the Poynting vector measures the flow of electric energy per unit time per unit area held perpendicular to the direction of propagation of the electromagnetic wave. It is also called 'flux vector' or 'power flux'.

The Poynting theorem in electrodynamics gives the conservation of energy in the form of a partial differential equation for the electromagnetic field. In classical mechanics, it is analogous to the work energy theorem and mathematically similar to the continuity equation, since it relates the stored energy in the electromagnetic field to that of the work done on a charge distribution through energy flux. In other words, it states that the rate of work done on the charges in volume V enclosed by surface S by the Lorentz force is equal to the rate of decrease of energy stored in electromagnetic field minus the rate at which energy flows out of the surface S.

Mathematically,

$$\frac{dW}{dt} = -\frac{\partial}{\partial t}\iiint_V \frac{1}{2}\left(\varepsilon_0 \vec{E^2} + \frac{1}{\mu_0}\vec{B^2}\right)dV - \frac{1}{\mu_0}\oiint_S \left(\vec{E} \times \vec{B}\right)\cdot \vec{ds}$$

Let us consider the current distribution in a volume V and some charges enclosed by a surface S. Let the surface enclose a produced electric field $\vec{E}$ and magnetic field $\vec{B}$. If charges move slightly through a distance dl in small time dt, then the work is done by the Lorentz force (electromagnetic force) acting on these charges.

Electromagnetic force acting on a charge dq is given by,

$$\vec{F} = dq\left[\vec{E} + \left(\vec{V} \times \vec{B}\right)\right]$$

Work done by the electromagnetic force in displacing charge dq through $\vec{dl}$ is given by

$$dW = \vec{F}\cdot\vec{dl} = dq\left[\vec{E} + \left(\vec{V} \times \vec{B}\right)\right]\cdot\vec{dl} \tag{2.60}$$

If $\vec{F}$ acts for small time dt on the charge, then velocity of charge is given by,

$$\vec{v} = \frac{\vec{dl}}{dt}$$

$$\text{or } \vec{dl} = \vec{v}dt \tag{2.61}$$

Substituting the value of $\vec{dl}$ from Equation (2.61) in Equation (2.60), we get

$$dW = dq\left[\vec{E} + \left(\vec{V} \times \vec{B}\right)\right] \cdot \vec{v}dt$$

$$\text{or } dW = \left[\vec{E}dq + \left(\vec{V} \times \vec{B}\right)dq\right] \cdot \vec{v}dt$$

$$\text{or } dW = \left(\vec{E} \cdot \vec{v}\right)dqdt + \left(\vec{V} \times \vec{B}\right) \cdot \vec{v}dqdt \qquad (2.62)$$

But, $\left(\vec{v} \times \vec{B}\right) \cdot \vec{v} = 0$, because $\vec{v}$ and $\left(\vec{v} \times \vec{B}\right)$ are perpendicular
Therefore, Equation (2.62) becomes,

$$dW = \left(\vec{E} \cdot \vec{v}\right)dqdt + 0$$

or

$$dW = \left(\vec{E} \cdot \vec{v}\right)dqdt \qquad (2.63)$$

If ρ = volume charge density, then

$$\rho = \frac{dq}{dV}$$

$$\text{or } dq = \rho dV$$

Therefore, Equation (2.63) becomes,

$$dW = \left(\vec{E} \cdot \vec{v}\right)\rho dVdt$$

$$\text{or } dW = \left(\vec{E} \cdot \vec{v}\rho\right)dVdt$$

But $\vec{v}\rho = \vec{J}$
Therefore, $dW = \left(\vec{E} \cdot \vec{J}\right)dVdt$

$$\text{or } \frac{dW}{dt} = \left(\vec{E} \cdot \vec{J}\right)dV$$

Thus, the rate at which the total work is done on all the charges in volume V is given by

$$\frac{dW}{dt} = \iiint_V \left(\vec{E} \cdot \vec{J}\right)dV \qquad (2.64)$$

where, $\vec{E}\cdot\vec{J}$ = work done per unit time per unit volume by $\vec{E}$; however, work done by $\vec{B}$ is zero.

As we know, the fourth wave equation of Maxwell is given by,

$$\vec{\nabla}\times\vec{B} = \mu_0\left(\vec{J}+\varepsilon_0\frac{\partial\vec{E}}{\partial t}\right)$$

$$\text{or } \frac{1}{\mu_0}\left(\vec{\nabla}\times\vec{B}\right) = \vec{J}+\varepsilon_0\frac{\partial\vec{E}}{\partial t}$$

$$\text{or } \vec{J} = \frac{1}{\mu_0}\left(\vec{\nabla}\times\vec{B}\right)-\varepsilon_0\frac{\partial\vec{E}}{\partial t}$$

Therefore,

$$\vec{E}\cdot\vec{J} = \frac{1}{\mu_0}\vec{E}\cdot\left(\vec{\nabla}\times\vec{B}\right)-\varepsilon_0\vec{E}\cdot\frac{\partial\vec{E}}{\partial t} \tag{2.65}$$

$$\text{Since, } \vec{\nabla}\cdot\left(\vec{A}\times\vec{B}\right) = \vec{B}\cdot\left(\vec{\nabla}\times\vec{A}\right)-\vec{A}\cdot\left(\vec{\nabla}\times\vec{B}\right)$$

$$\text{or } \vec{A}\cdot\left(\vec{\nabla}\times\vec{B}\right) = \vec{B}\cdot\left(\vec{\nabla}\times\vec{A}\right)-\vec{\nabla}\cdot\left(\vec{A}\times\vec{B}\right) \tag{2.66}$$

With the help of Equation (2.66), Equation (2.65) can be written as,

$$\vec{E}\cdot\vec{J} = \frac{1}{\mu_0}\left[\vec{B}\cdot\left(\vec{\nabla}\times\vec{A}\right)-\vec{\nabla}\cdot\left(\vec{E}\times\vec{B}\right)\right]-\varepsilon_0\vec{E}\frac{\partial\vec{E}}{\partial t} \tag{2.67}$$

According to Faraday's law of electromagnetic induction,

$$\vec{\nabla}\times\vec{E} = -\frac{\partial\vec{B}}{\partial t}$$

Therefore, Equation (2.67) can be written as,

$$\vec{E}\cdot\vec{J} = \frac{1}{\mu_0}\left[\vec{B}\cdot\left(-\frac{\partial\vec{B}}{\partial t}\right)-\vec{\nabla}\left(\vec{E}\times\vec{B}\right)\right]-\varepsilon_0\vec{E}\frac{\partial\vec{E}}{\partial t}$$

$$\text{or } \vec{E}\cdot\vec{J} = \frac{1}{2\mu_0}2\vec{B}\cdot\frac{\partial\vec{B}}{\partial t}-\frac{1}{2}\varepsilon_0\vec{E}\cdot\frac{\partial\vec{E}}{\partial t}-\frac{1}{\mu_0}\vec{\nabla}\cdot\left(\vec{E}\times\vec{B}\right) \tag{2.68}$$

$$\text{But, } \frac{\partial}{\partial t}\left(\vec{E}\cdot\vec{E}\right) = \vec{E}\cdot\frac{\partial\vec{E}}{\partial t}+\vec{E}\cdot\frac{\partial\vec{E}}{\partial t} = 2\vec{E}\cdot\frac{\partial\vec{E}}{\partial t}$$

$$\text{Or, } \frac{\partial}{\partial t}\left(\vec{E}\cdot\vec{E}\right) = 2\vec{E}\cdot\frac{\partial\vec{E}}{\partial t}$$

$$\text{Also } \frac{\partial}{\partial t}\left(\vec{B}\cdot\vec{B}\right) = 2\vec{B}\cdot\frac{\partial\vec{B}}{\partial t}$$

Therefore, Equation (2.68) becomes,

$$\vec{E}\cdot\vec{J} = -\frac{1}{2\mu_0}\frac{\partial\left(\vec{B}\cdot\vec{B}\right)}{\partial t} - \frac{1}{2}\varepsilon_0\frac{\partial\left(\vec{E}\cdot\vec{E}\right)}{\partial t} - \frac{1}{\mu_0}\vec{\nabla}\times\left(\vec{E}\cdot\vec{B}\right)$$

$$\text{or } \vec{E}\cdot\vec{J} = -\frac{1}{2}\left(\frac{\varepsilon_0\partial\overrightarrow{E^2}}{\partial t} + \frac{1}{\mu_0}\frac{\partial\overrightarrow{B^2}}{\partial t}\right) - \frac{1}{\mu_0}\vec{\nabla}\times\left(\vec{E}\cdot\vec{B}\right)$$

Substituting $\vec{E}\cdot\vec{J}$ in Equation (2.68), we get

$$\frac{dW}{dt} = \iiint_V \left[-\frac{1}{2}\left(\frac{\varepsilon_0\partial\overrightarrow{E^2}}{\partial t} + \frac{1}{\mu_0}\frac{\partial\overrightarrow{B^2}}{\partial t}\right) - \frac{1}{\mu_0}\vec{\nabla}\times\left(\vec{E}\cdot\vec{B}\right)\right]dV$$

$$\text{or } \frac{dW}{dt} = -\frac{\partial}{\partial t}\iiint_V \frac{1}{2}\left(\varepsilon_0\overrightarrow{E^2} + \frac{1}{\mu_0}\overrightarrow{B^2}\right)dV - \frac{1}{\mu_0}\iiint_V \vec{\nabla}\left(\vec{E}\times\vec{B}\right)dV \tag{2.69}$$

According to Gauss's divergence theorem,

$$\iiint_V \vec{\nabla}\cdot\left(\vec{E}\times\vec{B}\right)dV = \oiint_S \left(\vec{E}\times\vec{B}\right)\cdot\vec{dS}$$

Therefore, Equation (2.69) becomes,

$$\frac{dW}{dt} = -\frac{\partial}{\partial t}\iiint_V \frac{1}{2}\left(\varepsilon_0\overrightarrow{E^2} + \frac{1}{\mu_0}\overrightarrow{B^2}\right)dV - \frac{1}{\mu_0}\oiint_S \left(\vec{E}\times\vec{B}\right)\cdot\vec{dS}$$

This is the required expression for the Poynting theorem.

In this expression $\frac{dW}{dt}$ represents the rate at which work is done on all the charges in volume V, $\iiint_V \frac{1}{2}\left(\varepsilon_0\overrightarrow{E^2} + \frac{1}{\mu_0}\overrightarrow{B^2}\right)dV = \mathrm{U_{em}}$, represents the total energy stored in the electric and magnetic fields and $\frac{1}{\mu_0}\oiint_S \left(\vec{E}\times\vec{B}\right)\cdot\vec{dS}$ represents the rate at which energy flows out of the surface S enclosing the volume V.

2.10 ELECTROMAGNETIC WAVE PROPAGATION THROUGH VACUUM

Maxwell's equations in a medium are given as,

$$\vec{\nabla}\cdot\vec{E}=\frac{\rho}{\varepsilon_0}$$

$$\vec{\nabla}\cdot\vec{B}=0$$

$$\vec{\nabla}\times\vec{E}=-\frac{\partial\vec{B}}{\partial t}$$

$$\vec{\nabla}\times\vec{B}=\mu\vec{J}+\mu\varepsilon\frac{\partial\vec{E}}{\partial t}$$

For vacuum or free space, charge density $\rho = 0$, current density $\vec{J} = 0$ and $\mu = \mu_0$, $\varepsilon = \varepsilon_0$.

Therefore, the above equations can be written as,

$$\vec{\nabla}\cdot\vec{E}=0$$

$$\vec{\nabla}\cdot\vec{B}=0$$

$$\vec{\nabla}\times\vec{E}=-\frac{\partial\vec{B}}{\partial t}$$

$$\text{and } \vec{\nabla}\times\vec{B}=\mu_0\varepsilon_0\frac{\partial\vec{E}}{\partial t}$$

Case 1: Wave equation for electric field vector $\vec{\mathbf{E}}$

We know that, $\vec{\nabla}\times\vec{E}=-\frac{\partial\vec{B}}{\partial t}$

Taking curl on both sides, we get

$$\vec{\nabla}\times\left(\vec{\nabla}\times\vec{E}\right)=-\vec{\nabla}\times\frac{\partial\vec{B}}{\partial t} \tag{2.70}$$

Applying the triple vector identity, $\vec{A}\times\left(\vec{B}\times\vec{C}\right)=\vec{B}\left(\vec{A}\cdot\vec{C}\right)-\vec{C}\left(\vec{A}\cdot\vec{B}\right)$, we get

$$\vec{\nabla}\left(\vec{\nabla}\cdot\vec{E}\right)-\vec{E}\left(\vec{\nabla}\cdot\vec{\nabla}\right)=-\vec{\nabla}\times\frac{\partial\vec{B}}{\partial t}$$

$$\text{or } \vec{\nabla}\left(\vec{\nabla}\cdot\vec{E}\right)-\nabla^2\vec{E}=-\vec{\nabla}\times\frac{\partial\vec{B}}{\partial t}$$

$$\text{or } \vec{\nabla}\left(\vec{\nabla}\cdot\vec{E}\right)-\vec{\nabla}^2\vec{E}=-\frac{\partial}{\partial t}\left(\vec{\nabla}\times\vec{B}\right) \tag{2.71}$$

But from Equations (2.70) and (2.71), we have

$$\vec{\nabla}\cdot\vec{E}=0 \text{ and } \vec{\nabla}\times\vec{B}=\mu_0\varepsilon_0\frac{\partial\vec{E}}{\partial t}$$

Therefore, Equation (2.71) becomes,

$$0-\nabla^2\vec{E}=-\frac{\partial}{\partial t}\left(\mu_0\varepsilon_0\frac{\partial\vec{E}}{\partial t}\right)$$

$$\text{or } \nabla^2\vec{E}=\mu_0\varepsilon_0\frac{\partial^2\vec{E}}{\partial t^2} \tag{2.72}$$

Thus, Equation (2.72) represents the wave equation for electric field vector $\vec{E}$ in free space or in vacuum.

Case 2: Wave equation for magnetic field vector $\vec{B}$

Since, $\vec{\nabla}\times\vec{B}=\mu_0\varepsilon_0\frac{\partial\vec{E}}{\partial t}$

Taking curl on both sides, we get

$$\vec{\nabla}\times\left(\vec{\nabla}\times\vec{B}\right)=\vec{\nabla}\times\left(\mu_0\varepsilon_0\frac{\partial\vec{E}}{\partial t}\right)$$

Applying the triple vector identity, $\vec{A}\times\left(\vec{B}\times\vec{C}\right)=\vec{B}\left(\vec{A}\cdot\vec{C}\right)-\vec{C}\left(\vec{A}\cdot\vec{B}\right)$, we get

$$\vec{\nabla}\left(\vec{\nabla}\cdot\vec{B}\right)-\vec{B}\left(\vec{\nabla}\cdot\vec{\nabla}\right)=\vec{\nabla}\times\left(\mu_0\varepsilon_0\frac{\partial\vec{E}}{\partial t}\right)$$

$$\text{or } \vec{\nabla}\left(\vec{\nabla}\cdot\vec{B}\right)-\nabla^2\vec{B}=\mu_0\varepsilon_0\frac{\partial}{\partial t}\left(\vec{\nabla}\times\vec{E}\right) \tag{2.73}$$

From Equations (2.72) and (2.73), we have

$$\vec{\nabla}\cdot\vec{B}=0 \text{ and } \vec{\nabla}\times\vec{E}=-\frac{\partial\vec{B}}{\partial t}$$

Therefore, Equation (2.73) can be written as,

$$0-\nabla^2\vec{B}=-\mu_0\varepsilon_0\frac{\partial}{\partial t}\left(\frac{\partial\vec{B}}{\partial t}\right)$$

$$\text{or } \nabla^2\vec{B} = \mu_0\varepsilon_0 \frac{\partial^2\vec{B}}{\partial t^2} \tag{2.74}$$

Equation (2.74) represents the wave equation for the vector field $\vec{B}$ in free space or in vacuum.

Hence, wave equation for $\vec{E}$ and $\vec{B}$ in free space or vacuum is given as,

$$\nabla^2\vec{E} = \mu_0\varepsilon_0 \frac{\partial^2\vec{E}}{\partial t^2}$$

$$\text{and } \nabla^2\vec{B} = \mu_0\varepsilon_0 \frac{\partial^2\vec{B}}{\partial t^2}$$

2.11 ENERGY DENSITY IN ELECTROMAGNETIC FIELD

The electromagnetic field is the sum of an electric field and a magnetic field. Therefore, energy density in an electromagnetic field, u = energy density in an electric field u_E + energy density in a magnetic field u_B,

$$u = u_E + u_B \tag{2.75}$$

Now, as we know, the energy density in an electric field is given as,

$$u_E = \frac{1}{2}\varepsilon_0 E^2 \tag{2.76}$$

and energy density in a magnetic field is given as,

$$u_B = \frac{1}{2}\frac{B^2}{\mu_0} \tag{2.77}$$

Using Equations (2.76) and (2.77) in Equation (2.75), we get

$$u = \frac{1}{2}\left(\varepsilon_0 E^2 + \frac{B^2}{\mu_0}\right)$$

Thus, the above equation gives the energy density in the electromagnetic field.

2.12 ELECTROMAGNETIC WAVE PROPAGATION THROUGH ISOTROPIC DIELECTRIC MEDIUM

An isotropic dielectric medium is the one whose polarization has a direction that is parallel to the applied electric field and a magnitude which does not depend on the direction of an electric field. Let $\vec{E}$ and $\vec{B}$ be the electric and magnetic fields at any point in a medium having permeability μ and permittivity ε, then the Maxwell equations in the medium are given as,

$$\vec{\nabla} \times \vec{E} = \frac{\rho_f}{\varepsilon_0} \tag{2.78}$$

$$\vec{\nabla} \cdot \vec{B} = 0 \tag{2.79}$$

$$\vec{\nabla} \times \vec{E} = -\frac{\partial \vec{B}}{\partial t} \tag{2.80}$$

$$\text{and } \vec{\nabla} \times \vec{B} = \mu \vec{J}_f + \mu\varepsilon \frac{\partial \vec{E}}{\partial t} \tag{2.81}$$

Case 1: Wave equation for electric field vector $\vec{E}$

$$\text{Since, } \vec{\nabla} \times \vec{E} = -\frac{\partial \vec{B}}{\partial t}$$

Taking curl on both sides, we get

$$\vec{\nabla} \times \left(\vec{\nabla} \times \vec{E}\right) = -\vec{\nabla} \times \left(\frac{\partial \vec{B}}{\partial t}\right) \tag{2.82}$$

Applying the triple vector identity, $\vec{A} \times \left(\vec{B} \times \vec{C}\right) = \vec{B}\left(\vec{A} \cdot \vec{C}\right) - \vec{C}\left(\vec{A} \cdot \vec{B}\right)$, we get

$$\vec{\nabla}\left(\vec{\nabla} \cdot \vec{E}\right) - \vec{E}\left(\vec{\nabla} \cdot \vec{\nabla}\right) = -\vec{\nabla} \times \frac{\partial \vec{B}}{\partial t}$$

$$\text{or } \vec{\nabla}\left(\vec{\nabla} \cdot \vec{E}\right) - \vec{\nabla}^2 \vec{E} = -\frac{\partial}{\partial t}\left(\vec{\nabla} \times \vec{B}\right) \tag{2.83}$$

Using Equations (2.78) and (2.81) in Equation (2.83), we get

$$\vec{\nabla}\left(\frac{\rho_f}{\varepsilon_0}\right) - \nabla^2 \vec{E} = -\frac{\partial}{\partial t}\left(\mu \vec{J}_f + \mu\varepsilon \frac{\partial \vec{E}}{\partial t}\right)$$

$$\text{or } \vec{\nabla}\left(\frac{\rho_f}{\varepsilon_0}\right) - \nabla^2 \vec{E} = -\frac{\partial}{\partial t}\left(\mu \vec{J}_f\right) - \mu\varepsilon \frac{\partial^2 \vec{E}}{\partial t^2} \tag{2.84}$$

If there is no free charge and no current in the dielectric medium, then $\rho_f = 0$ and $\vec{J}_f = 0$

Therefore Equation (2.84) can be written as,

$$0-\nabla^2\vec{E}=0-\mu\varepsilon\frac{\partial^2\vec{E}}{\partial t^2}$$

$$\text{or}\ \nabla^2\vec{E}=\mu\varepsilon\frac{\partial^2\vec{E}}{\partial t^2} \tag{2.85}$$

This is the electromagnetic wave equation in terms of electric field vector $\vec{E}$ in the dielectric medium.

Case 2: Wave equation for the magnetic field vector $\vec{B}$

$$\text{Since, } \vec{\nabla}\times\vec{B}=\mu\vec{J}_f+\mu\varepsilon\frac{\partial^2\vec{E}}{\partial t^2}$$

Taking curl on both sides, we get

$$\vec{\nabla}\times\left(\vec{\nabla}\times\vec{B}\right)=\vec{\nabla}\times\left(\mu\vec{J}_f+\mu\varepsilon\frac{\partial\vec{E}}{\partial t}\right)$$

$$\text{or}\ \vec{\nabla}\times\left(\vec{\nabla}\times\vec{B}\right)=\vec{\nabla}\times\left(\mu\vec{J}_f\right)+\mu\varepsilon\left(\vec{\nabla}\times\frac{\partial\vec{E}}{\partial t}\right) \tag{2.86}$$

Applying the triple vector identity, $\vec{A}\times\left(\vec{B}\times\vec{C}\right)=\vec{B}\left(\vec{A}\cdot\vec{C}\right)-\vec{C}\left(\vec{A}\cdot\vec{B}\right)$, we get

$$\vec{\nabla}\left(\vec{\nabla}\cdot\vec{B}\right)-\vec{B}\left(\vec{\nabla}\cdot\vec{\nabla}\right)=\mu\left(\vec{\nabla}\times\vec{J}_f\right)+\mu\varepsilon\left(\vec{\nabla}\times\frac{\partial\vec{E}}{\partial t}\right)$$

Since, $\vec{\nabla}\cdot\vec{B}=0$ and $\vec{\nabla}\cdot\vec{\nabla}=\nabla^2$; therefore, the above equation becomes,

$$0-\vec{\nabla}^2\vec{B}=\mu\left(\vec{\nabla}\times\vec{J}_f\right)+\mu\varepsilon\frac{\partial}{\partial t}\left(\vec{\nabla}\times\vec{E}\right)$$

$$\text{or}\ -\nabla^2\vec{B}=\mu\left(\vec{\nabla}\times\vec{J}_f\right)+\mu\varepsilon\frac{\partial}{\partial t}\left(\vec{\nabla}\times\vec{E}\right) \tag{2.87}$$

Also, from Equation (2.83), we know that $\vec{\nabla}\times\vec{E}=-\frac{\partial\vec{B}}{\partial t}$,
Therefore, Equation (2.87) becomes,

$$-\nabla^2\vec{B}=\mu\left(\vec{\nabla}\times\vec{J}_f\right)+\mu\varepsilon\frac{\partial}{\partial t}\left(-\frac{\partial\vec{B}}{\partial t}\right)$$

$$\text{or}\ -\nabla^2\vec{B}=\mu\left(\vec{\nabla}\times\vec{J}_f\right)-\mu\varepsilon\frac{\partial^2\vec{B}}{\partial t^2} \tag{2.88}$$

For dielectric medium $\vec{J}_f = 0$,
Therefore, Equation (2.88) can be written as,

$$-\nabla^2 \vec{B} = 0 - \mu\varepsilon \frac{\partial^2 \vec{B}}{\partial t^2}$$

$$\text{or } \nabla^2 \vec{B} = \mu\varepsilon \frac{\partial^2 \vec{B}}{\partial t^2} \tag{2.89}$$

This is the electromagnetic wave equation in terms of magnetic field vector $\vec{B}$ in the dielectric medium.

2.13 TRANSVERSE NATURE OF ELECTROMAGNETIC WAVES

A mechanical wave is said to be a transverse wave if the particles of the medium vibrate perpendicular to the direction of propagation of the wave in the medium. However, in an electromagnetic wave, the oscillating electric field and magnetic field vectors are restricted in a plane perpendicular to the direction of propagation of the electromagnetic wave.

In the first method, let us consider a plane electromagnetic wave that is travelling along X-direction, then the oscillating electric field and magnetic field vectors are restricted in Y- and Z-plane only. In a non-conducting and charge-free medium, Maxwell's equations are written as,

$$\vec{\nabla} \cdot \vec{E} = 0$$

$$\vec{\nabla} \cdot \vec{B} = 0p$$

$$\vec{\nabla} \times \vec{E} = -\frac{\partial \vec{B}}{\partial t}$$

$$\text{and } \vec{\nabla} \times \vec{B} = \mu\varepsilon \frac{\partial \vec{E}}{\partial t}$$

Now equations of electromagnetic waves in terms of electric field vector $\vec{E}$ and magnetic field vector $\vec{B}$ are given by,

$$\nabla^2 \vec{E} = \mu\varepsilon \frac{\partial^2 \vec{E}}{\partial t^2} \tag{2.90}$$

$$\text{and } \nabla^2 \vec{B} = \mu\varepsilon \frac{\partial^2 \vec{E}}{\partial t^2} \tag{2.91}$$

The solutions of wave Equations (2.90) and (2.91) are given by,

$$\vec{E} = \vec{E}_0 e^{i(\omega t - \vec{k}\cdot\vec{r})} \tag{2.92}$$

$$\text{and } \vec{B} = \vec{B}_0 e^{i(\omega t - \vec{k}\cdot\vec{r})} \tag{2.93}$$

where, $\vec{k}$ is the propagation vector, ω is angular frequency and $\vec{r}$ is the projection of position vector $\vec{r}$.

Since $\vec{\nabla}\cdot\vec{E} = 0$,

Therefore, $$\vec{\nabla}\cdot\vec{E}_0 e^{i(\omega t - \vec{k}\cdot\vec{r})} = 0$$

$$\text{or } \vec{k}\cdot\vec{E} = 0 \tag{2.94}$$

Hence, electric field vector $\vec{E}$ is perpendicular to the propagation vector $\vec{k}$.

Now substituting Equations (2.92) and (2.93) in $\vec{\nabla}\times\vec{E} = -\frac{\partial\vec{B}}{\partial t}$, we get

$$\vec{\nabla}\times\vec{E}_0 e^{i(\omega t - \vec{k}\cdot\vec{r})} = -\frac{\partial}{\partial t}\vec{B}_0 e^{i(\omega t - \vec{k}\cdot\vec{r})}$$

$$\text{or } \vec{E}_0 \times \left(-i\vec{k}\right) e^{i(\omega t - \vec{k}\cdot\vec{r})} = -\vec{B}_0 \times (i\omega) e^{i(\omega t - \vec{k}\cdot\vec{r})}$$

$$\text{or } \vec{k}\times\vec{E} = \omega\vec{B} \tag{2.95}$$

This equation shows that magnetic field $\vec{B}$ is perpendicular to both $\vec{k}$ and $\vec{E}$.

Since electric field vector $\vec{E}$ is perpendicular to $\vec{k}$, so $\vec{B}$ is also perpendicular to $\vec{E}$. Thus $\vec{k}$, $\vec{E}$ and $\vec{B}$ are mutually perpendicular to each other. These three vectors $\vec{k}$, $\vec{E}$ and $\vec{B}$ are as shown in Figure 2.6. Since both electric field vector $\vec{E}$ and magnetic field vector $\vec{B}$ are perpendicular to $\vec{k}$, electromagnetic wave is transverse in nature.

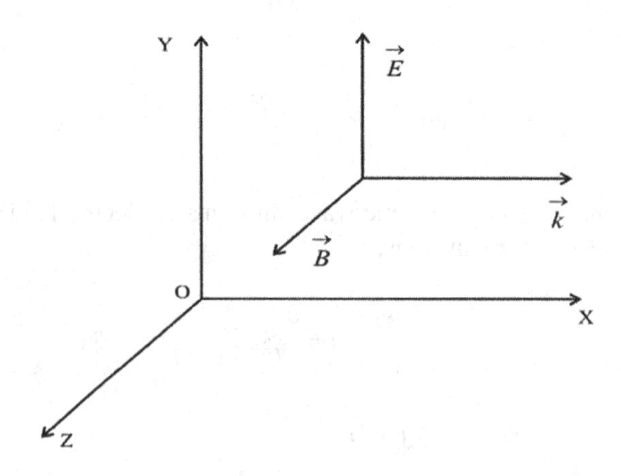

FIGURE 2.6 Three vector fields perpendicular to each other.

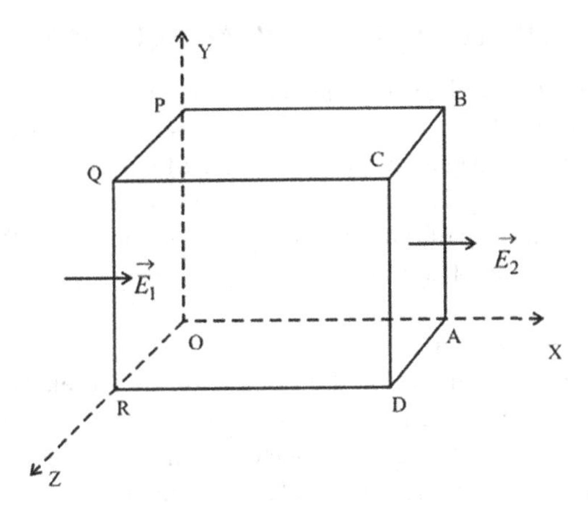

FIGURE 2.7 Cube with sides parallel to the coordinate planes.

In a second method, let us suppose a plane electromagnetic wave is moving in the *+ve* x direction, consider a cube with sides parallel to the coordinate planes as shown in Figure 2.7. We are considering, for this situation, the electric flux or magnetic flux through this cube is zero.

According to Gauss's law

$$\oint \vec{E} \cdot \vec{dS} = 0 \tag{2.96}$$

Since it is a plane wave, $\vec{E}$ depends only on x and t. Therefore, there is no contribution to net electric flux from the top and bottom or from the front and back faces of the cube.

Also, the electric flux through the left face of the cube is $E_1 \times$ area OPQR, where E_1 is the electric field at the face OPQR. The electric flux out of the right face is $E_2 \times$ area ABCD, where E_2 is the electric field at face ABCD.

Therefore, Equation (2.96) becomes,

$$E_1 \times \text{area OPQR} - E_2 \times \text{area ABCD} = 0$$

$$\text{or } A(E_1 - E_2) = 0$$

Since 'area' cannot be zero.

Therefore, $E_1 - E_2 = 0$

Two possibilities arise, either $E_1 = E_2$ or $E_1 = E_2 = 0$. In case $E_1 = E_2$, it would mean that an electric field associated with an electromagnetic wave is constant. But such a constant field cannot give rise to a wave. Therefore, $E_1 = E_2 = 0$, i.e., there is no electric field in the direction of propagation of the wave. This means that the electric field is perpendicular to the direction of propagation of the electromagnetic wave. A similar argument holds true for the magnetic field.

Since both electric field and magnetic field are perpendicular to the direction of propagation of the wave, electromagnetic waves are transverse in nature.

2.14 POLARIZATION OF ELECTROMAGNETIC WAVE

Polarization is described as the phenomenon of confining the vibration of an electric field vector $\vec{E}$ in a specific direction in any plane perpendicular to the direction of propagation of an electromagnetic wave. The state of polarization of an electromagnetic wave is designated by the geometrical shape which the tip of an electric field vector $\vec{E}$ draws as a function of time at a given point in the space. There, polarization is of three types as described below:

a) Linear or plane polarization
In a linear or plane polarization, if the orientation of the electric field vector $\vec{E}$ remains fixed as the electromagnetic wave travels, then the electromagnetic wave is said to be linear or plane polarized. The plane containing the electric field vector $\vec{E}$ and the direction of propagation of the electromagnetic wave is called 'plane of polarization'. The electric field vector $\vec{E}$ representing sinusoidal wave propagating along the X-direction is given by,

$$\vec{E} = \vec{E}_0 e^{i(\omega t - \vec{k}\cdot\vec{x})} = \vec{E}_0 e^{i(\omega t - kx)}$$

$$\text{or } \vec{E} = \vec{E}_0 \left(\cos(\omega t - kx) + i\sin(\omega t - kx)\right) \tag{2.97}$$

The real part of the electric field vector $\vec{E}$ is given by

$$E = E_0 \cos(\omega t - kx) \tag{2.98}$$

Now, consider two electromagnetic waves travelling along the X-axis. Suppose the electric field vector $\vec{E}$ of one wave is along the Y-axis, and the electric field vector $\vec{E}$

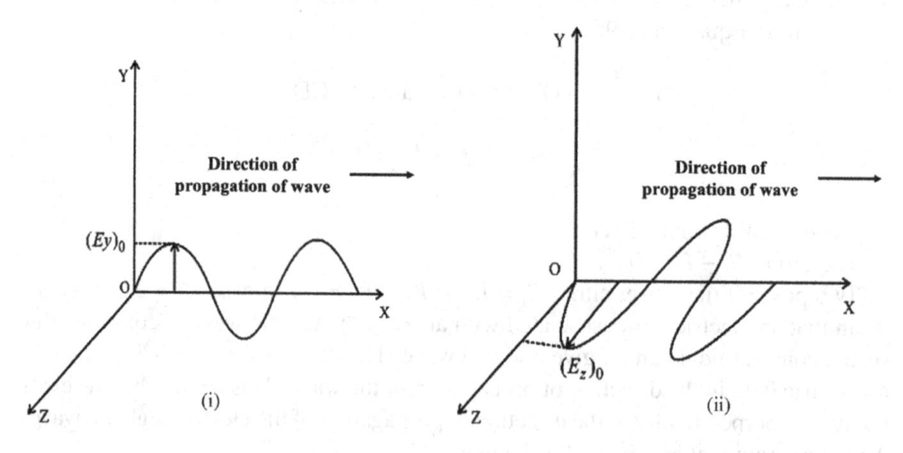

FIGURE 2.8 The propagation of wave with (i) vertical polarization and (ii) horizontal polarization.

of second wave is along the Z-axis, as shown in Figure 2.8(i) and Figure 2.8(ii), respectively.

The figures show that the wave is a polarized wave because, in both cases, the alteration of an electric field vector $\vec{E}$ is in a fixed plane perpendicular to the direction of propagation of the wave. Figure 2.8(i) presents the wave with a vertical polarization, and Figure 2.8(ii) presents the wave with a horizontal polarization.

The electric field vectors along the direction of Y- and Z-axis are given by,

$$E_y = (E_y)_0 \cos(\omega t - kx)$$

$$\text{or } \frac{E_y}{(E_y)_0} = \cos(\omega t - kx) \tag{2.99}$$

$$\text{and } E_z = (E_z)_0 \cos(\omega t - kx)$$

$$\text{or } \frac{E_z}{(E_z)_0} = \cos(\omega t - kx) \tag{2.100}$$

From Equations (2.99) and (2.100), we have

$$\frac{E_y}{(E_y)_0} = \frac{E_z}{(E_z)_0}$$

$$\text{or } E_y = \frac{(E_y)_0}{(E_z)_0} E_z \tag{2.101}$$

This represents the equation of a straight line with a slope of $\frac{(E_y)_0}{(E_z)_0}$. Hence, the tip of the resultant electric field vector $\vec{E}$ $\left(E = \sqrt{E_y^2 + E_z^2}\right)$ presents a straight line with a slope of $\frac{(E_y)_0}{(E_z)_0}$, as shown in Figure 2.9.

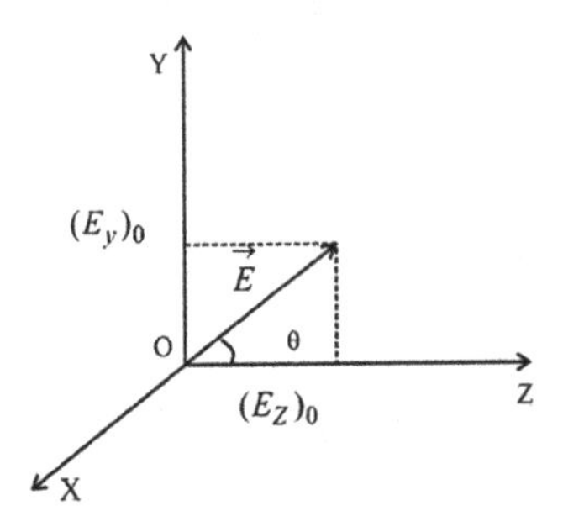

FIGURE 2.9. Resultant field vector in linear polarization representing a straight line.

Thus, the wave resulted from the superposition of two waves presented by E_y and E_z is linearly or plane polarized, and the polarization of the wave is known as 'linear' or 'plane polarization'.

b) Elliptical polarization

In the elliptical polarization of electromagnetic radiation, an ellipse is drawn on a fixed plane, such that the plane is normal and intersecting the direction of propagation. The tip of the electric field vector describes an ellipse in any fixed plane intersecting and normal to the direction of propagation. In this polarization, consider two electromagnetic waves of same frequency with their electric field vectors along Y- and Z-directions, respectively. These waves which are along Y- and Z-directions are represented as,

$$E_y = \left(E_y\right)_0 \cos\left(\omega t - kx\right) \tag{2.102}$$

$$\text{and } E_z = \left(E_z\right)_0 \cos\left(\omega t - kx + \phi\right) \tag{2.103}$$

where, ϕ is the phase difference between two electromagnetic waves.

Now, at $x = 0$, Equations (2.102) and (2.103) can be written as,

$$E_y = \left(E_y\right)_0 \cos \omega t \tag{2.104}$$

$$\text{and } E_z = \left(E_z\right)_0 \cos\left(\omega t + \phi\right) \tag{2.105}$$

From Equation (2.104)

$$\frac{E_y}{\left(E_y\right)_0} = \cos \omega t \tag{2.106}$$

Also, $\sin \omega t = \sqrt{1 - \cos^2 \omega t}$

Now, using Equation (2.106), we get

$$\sin \omega t = \sqrt{1 - \frac{E_y^2}{\left(E_y\right)_0^2}} \tag{2.107}$$

From Equation (2.104), we have

$$\frac{E_z}{\left(E_z\right)_0} = \cos\left(\omega t + \phi\right)$$

$$\text{or } \frac{E_z}{\left(E_z\right)_0} = \cos \omega t \cos \phi - \sin \omega t \sin \phi$$

Also, using Equations (2.105) and (2.106) in Equation (2.107), we get

$$\frac{E_z}{(E_z)_0} = \frac{E_y}{(E_y)_0}\cos\phi - \sqrt{1 - \frac{E_y^2}{(E_y)_0^2}}\sin\phi$$

$$\text{or } \frac{E_y}{(E_y)_0}\cos\phi - \frac{E_z}{(E_z)_0} = \sqrt{1 - \frac{E_y^2}{(E_y)_0^2}}\sin\phi$$

Squaring on both sides, we get

$$\left(\frac{E_y}{(E_y)_0}\cos\phi - \frac{E_z}{(E_z)_0}\right)^2 = \left(\sqrt{1 - \frac{E_y^2}{(E_y)_0^2}}\right)^2 \sin^2\phi$$

$$\text{Therefore, } \left(\frac{E_y}{(E_y)_0}\cos\phi\right)^2 + \left(\frac{E_z}{(E_z)_0}\right)^2 - 2\frac{E_y}{(E_y)_0}\cos\phi \times \frac{E_z}{(E_z)_0} = \left(1 - \frac{E_y^2}{(E_y)_0^2}\right)\sin^2\phi$$

$$\text{or } \frac{E_y^2}{(E_y)_0^2}\cos^2\phi + \frac{E_z^2}{(E_z)_0^2} - \frac{2E_yE_z}{(E_y)_0(E_z)_0}\cos\phi = \left(1 - \frac{E_y^2}{(E_y)_0^2}\right)\sin^2\phi$$

$$\text{or } \frac{E_y^2}{(E_y)_0^2} + \frac{E_z^2}{(E_z)_0^2} - \frac{2E_yE_z}{(E_y)_0(E_z)_0}\cos\phi = \sin^2\phi \tag{2.108}$$

The above equation represents an ellipse. Therefore, the tip of the electric field vector $\vec{E}$ of the resultant electromagnetic wave describes an ellipse as shown in Figure 2.10. Hence, the electromagnetic wave is elliptically polarized, and the polarization of electromagnetic wave is called 'elliptical polarization'.

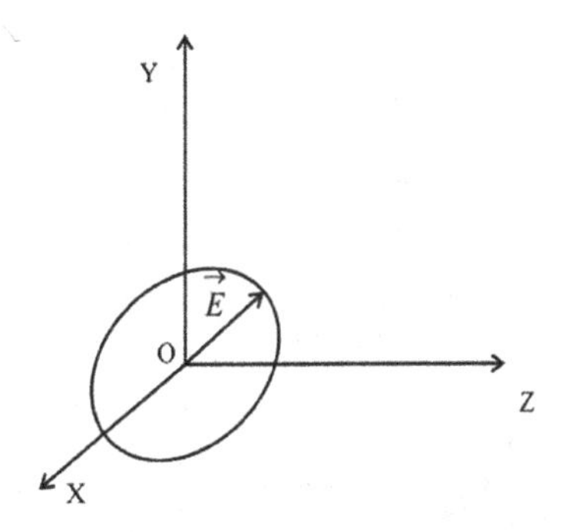

FIGURE 2.10 Resultant field vector in elliptical polarization representing an ellipse.

(c) Circular polarization

In the circular polarization of the electromagnetic wave, the magnitude of the electromagnetic wave is constant, whereas the direction rotates with a constant rate over a plane, such that the plane is perpendicular to the direction of propagation of wave. In this polarization, if the phase difference between two superposing electromagnetic waves is $\pi/2$ or 90° and the amplitude of both the waves is equal, i.e., $\phi = 90°$ and $(E_y)_0 = (E_z)_0 = E_0$, then Equation (2.108) is written as,

$$\frac{E_y^2}{(E_0)^2} + \frac{E_z^2}{(E_0)^2} - \frac{2E_yE_z}{E_0 \times E_0}\cos 90° = (\sin 90°)^2$$

$$\text{or}\ \frac{E_y^2}{(E_0)^2} + \frac{E_z^2}{(E_0)^2} - \frac{2E_yE_z}{E_0^2}\cos 90° = (\sin 90°)^2$$

$$\text{or}\ \frac{E_y^2}{E_0^2} + \frac{E_z^2}{E_0^2} - 0 = 1^2 \tag{2.109}$$

$$\text{or}\ E_y^2 + E_z^2 = E_0^2$$

This is the equation of a circle with radius equal to E_0. Therefore, the tip of the electric field of the resultant electromagnetic wave draws a circle of radius E_0 as shown in Figure 2.11.

Hence, the electromagnetic wave is circularly polarized and the polarization of the electromagnetic wave is known as 'circular polarization'. Now if the electric field vector $\vec{E}$ of the circularly polarized electromagnetic wave rotates anticlockwise, then the circularly polarized electromagnetic wave is known as 'left circularly polarized wave' as shown in Figure 2.12(i). On the other hand, if the electric field vector $\vec{E}$ of the circular polarized electromagnetic wave rotates clockwise, then the circularly polarized electromagnetic wave is known as 'right circularly polarized wave' as shown in Figure 2.12(ii).

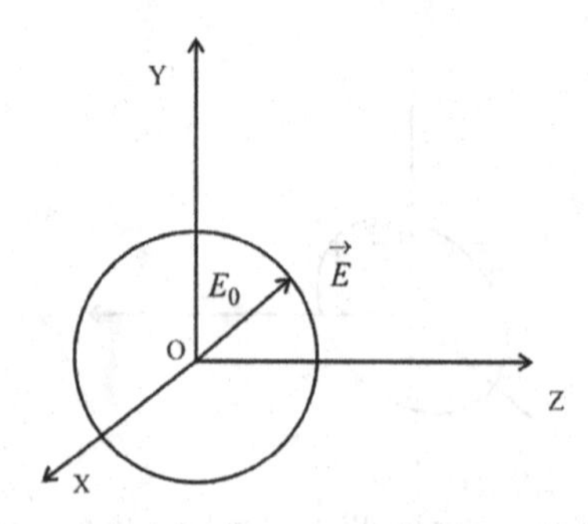

FIGURE 2.11 Resultant field vector in circular polarization representing a circle.

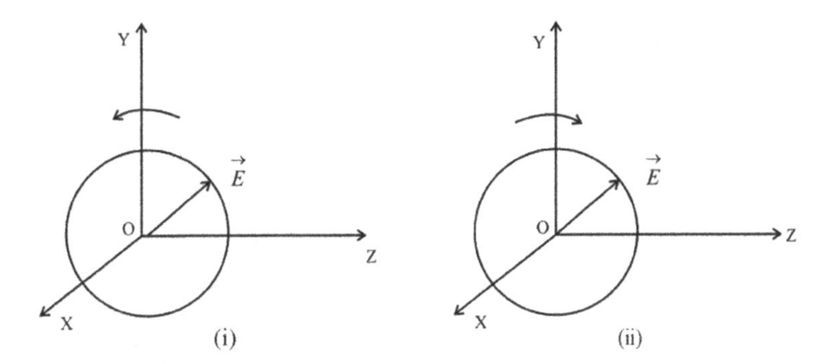

FIGURE 2.12 Resultant field vector in circular polarization representing a (i) left circularly polarized wave and (ii) right circularly polarized wave.

2.15 SUMMARY

In this chapter, Maxwell's equations were studied in different forms. Maxwell's equations are derived, which leads to the concept of electromagnetic waves. The equation of continuity proves the conservation of charge is possible. The concept of displacement current is explained, and its expression is derived followed by the Poynting vector and the Poynting theorem. A plane wave propagating in conductive material has amplitude attenuated by the exponential factor. By manipulation of Maxwell's equations, we were able to show that the propagation velocity of a light wave is governed by the electrical properties of the medium.

The transverse nature of electromagnetic wave is discussed in detail, and special cases were introduced in order to give thorough understanding of the propagation of the electromagnetic wave. Furthermore, this chapter described the polarization of an electromagnetic wave and its various types. In conclusion, the chapter highlights the properties and propagation of an electromagnetic wave in different mediums under imposed conditions.

FURTHER READINGS

1. A Dynamical Theory of the Electromagnetic Field. (1865). Maxwell's 1865 paper describing his 20 equations, link from Google Books.
2. Fleisch, D. (2008). *A Student's Guide to Maxwell's Equations*. Cambridge, UK: Cambridge University Press. ISBN 978-0-521-70147-1.
3. Imaeda, K. (1995). "Biquaternionic Formulation of Maxwell's Equations and their Solutions", in Ablamowicz, Rafał; Lounesto, Pertti (eds.), *Clifford Algebras and Spinor Structures*, Springer, pp. 265–280, doi:10.1007/978-94-015-8422-7_16, ISBN 978-90-481-4525-6
4. Reitz, J. R.; Milford, F. J.; Christy, R. W. (2008). "*Foundations of Electromagnetic Theory*" (4th ed.), Addison-Wesley Publishing Company, United States, ISBN 978-0-321-58174-7
5. Silagadze, Z. K. (2002). "Feynman's derivation of Maxwell equations and extra dimensions". *Annales de la Fondation Louis de Broglie* 27: 241–256. arXiv:hep-ph/0106235. Bibcode:2001hep.ph....6235S.
6. Weisstein, E. W. (1996–2007). "Poynting Theorem". From ScienceWorld – A Wolfram Web Resource.

3 Analog Communication Systems

3.1 INTRODUCTION

Electromagnetic (EM) waves are used in wireless communication. Wireless communication systems mainly use the principles of EM wave propagation in air, stratosphere, and ionosphere. Meanwhile, the other class of communication systems, i.e., optical fiber communication systems, use EM waves in the form of visible light spectrum in which, mainly, the propagation properties of the ray nature of EM wave is used.

The communication is in the form of electrical signals at the transmitter and receiver ends. At the transmitter end, these electrical signals (in the form of voltage or current) are converted into EM waves using an antenna. At the receiver end, EM waves are received by the receiver antenna and converted back to electrical signals. Thus, there is an important role of EM waves in communication systems.

Communication systems are broadly classified as (1) analog communication systems and (2) digital communication systems. The main difference between these two types of communication systems is the type of signal that has to be sent to the receiver from the transmitter. In analog communication systems, an analog signal is sent to the receiver, while a digital signal is transmitted in the case of digital communication systems. In this chapter, we will discuss analog communication systems. We will begin with a general introduction to communication systems. Further, we will discuss different signal processing steps which are required to convert the natural signal into suitable form for transmission. We will discuss commonly used modulation schemes, demodulation methods, and receiver designs applicable to analog communication systems.

3.2 COMMUNICATION SYSTEMS

Communication systems play an important role in sending the information generated by a source to the destination where it is required. A source can be anything like a human being generating speech, a musical instrument generating a sequence of tones, a sensor generating readings of physical quantity, etc. All signals generated by the sources are in analog form by default. So, we will discuss analog communication systems in this chapter. However, processing the signals in digital form has some advantages which will be discussed in subsequent chapters.

The role of a communication system is to process the message signal (signal to be transmitted) in such a way that it becomes suitable for transmission. Furthermore, the transmitted signal shall be regenerated/extracted at the receiver in its original form. Figure 3.1 represent a schematic of a simple communication system. A

DOI: 10.1201/9781003213468-3

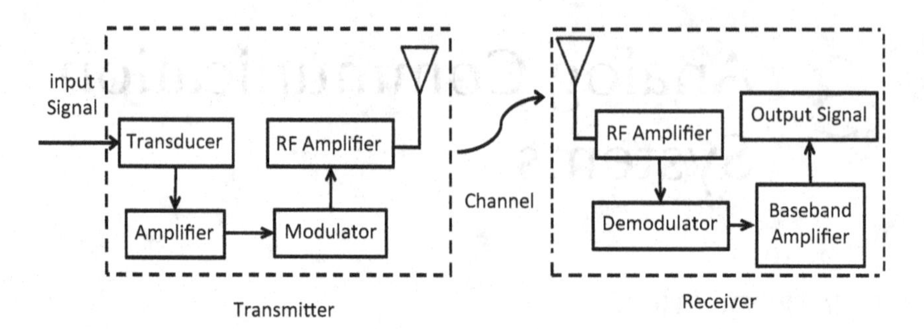

FIGURE 3.1 Block diagram of a transceiver system.

communication system comprises of three main modules: transmitter, receiver, and a channel. The structure and functionality of these modules are explained below.

3.2.1 Transmitter

A transmitter processes the signal that is generated by the source. This processing includes the following steps:

- Transformation of signal in electrical form using a suitable transducer, e.g., microphone to convert speech/music signal into electrical signal, camera to convert light intensity to electrical signal, etc.
- Amplification to improve the strength of the raw signal. The raw signal is also known as a 'baseband signal'.
- Modulation to translate the frequency to radio frequency (RF) range. The modulated signal is also known as a 'passband signal'.
- RF amplification to transmit the signal to the required distance.
- Conversion of RF signal into EM radiation using a suitable antenna.

3.2.2 Channel

A channel is the medium through which EM radiation propagates to reach the receiver (destination) where the information generated by the source has to be sent. The channel can be wired or wireless; however, we will be considering wireless channels in this book. The propagation of EM waves depends on the property of the medium. In wireless channels, the medium is mostly air except for satellite or ionospheric communication. In satellite communication, the signal has to propagate through different layers of the Earth's atmosphere and also through a vacuum, while in ionospheric communication, the signal propagates to the ionosphere (cloud of atmospheric gases in form of ions/plasma) and reflects back to the Earth. While the signal propagates from the transmitter to the receiver, the signal undergoes attenuation due to atmospheric losses. In free space, the attenuation can be quantified by the Friis equation which relates the transmitted power and the received power as

$$P_r = \frac{P_t G_t G_r \lambda^2}{4\pi d^2} \tag{3.1}$$

where P_t is the transmitted power, P_r is the received power, d is the distance between transmitter and receiver, G_t and G_r are the gains of transmitter antenna and the receiver antenna, respectively.

The signal undergoes various processes like reflection, refraction, diffraction, scattering, etc. while propagation from the transmitter to the receiver. In addition to these physical processes, noise gets added not only at the channel but also at the transmitter and receiver ends due to semiconductor devices involved in processing of the signals.

3.2.3 Receiver

A receiver processes the received signal and extracts the message signal. The processes involved at the receiver end are summarized in brief as follows:

- Conversion of EM signal to electrical signal using a suitable receiver antenna for further processing of the received signal.
- Filtering to remove unwanted frequencies from the received signal.
- RF amplification/low noise amplification to improve the received signal strength.
- Demodulation for frequency translation to baseband frequency range.
- Conversion of baseband electrical signal to a suitable form using a corresponding transducer, e.g., speaker for conversion to audio form, LED/LCD for display of images/videos, etc.

These are basic processes involved at various stages of communication systems. However, nowadays in advanced systems, further processing is included like analog to digital conversion for digital systems, stereo for quality enhancement of audio signals, cryptography for enhancing security of any signal, compression to reduce the amount of data that needs transmission, error correction/detection mechanisms, etc.

The important process during communication is modulation. We will discuss the need for modulation in the next section.

3.3 MODULATION

Modulation is superimposing the baseband signal over a high frequency signal such that the information can be extracted in its original form by appropriate signal processing techniques. Using modulation, one of the parameters, amplitude, phase, or frequency, of the high frequency signal is changed proportionate to the instantaneous value of the baseband message signal. In modulation, the message signal or the baseband signal is known as a ‘modulating signal’, and the high frequency signal is known as a 'carrier signal'. The process of modulation translates the frequency of baseband signal to the carrier frequency. The receiver is designed in accordance with the modulation scheme used at the transmitter. Accordingly, the modulation schemes are classified as:

1. Amplitude Modulation: The amplitude of a carrier signal is changed according to instantaneous value of the message signal.
2. Angle Modulation: Phase and frequency are part of angle in sinusoidal representation of signals. So, phase modulation and frequency modulation are classified as angle modulation scheme. In the latter part of this chapter, we will establish a relation between phase modulation and frequency modulation.

3.3.1 Need for Modulation

Most message signals generated naturally are low frequency signals, or the frequency spectrum of natural signals is around low frequencies and mostly centered at zero frequency. Transducers like microphones are used to convert these practical/natural signals into electrical form (voltage or current). Multiple such electrical signals are to be transmitted simultaneously. We consider wireless transmission media in this book. So, the following discussion justifies the need for modulation.

3.3.1.1 Multiplexing for Simultaneous Transmission of Signals

Multiple signals having similar frequency content are transmitted; for example, multiple simultaneous phone calls transmitting speech signals over a wireless medium. If these signals are transmitted in the raw form, there will be a lot of interference, and the receiver would not be able to bring back the received signal into useful form. Thus, through modulation, we can translate the different signals at different frequencies for transmission without interference among themselves. This is realized using carrier signals with different frequencies such that the signals do not overlap after modulation.

3.3.1.2 Practical Antenna Height/Dimension

The radiation from an antenna can be effective only if the antenna dimensions are of the order of one quarter of the signal wavelength to be transmitted. In case of a dipole antenna, the minimum length should be $\lambda/4$; for baseband signals, the wavelengths of the signals are usually very high, e.g., the frequency spectrum of a baseband speech signal ranges up to 3400 Hz; correspondingly, the antenna length/height becomes impractical. Using suitable frequency carrier for modulation, the required antenna length/height for effective radiation can be brought to practical values.

We will discuss these modulation schemes, circuits to realize these modulation schemes, and the corresponding demodulation circuitry in the following sections.

3.4 AMPLITUDE MODULATION

As stated earlier, the amplitude of the carrier signal is modulated using amplitude modulation (AM). There are many variations in AM based on the application and bandwidth/power considerations as follows:

1. Double sideband-suppressed carrier amplitude modulation
2. Amplitude modulation
3. Single sideband modulation
4. Vestigial sideband modulation

In this section, we will discuss various amplitude modulation schemes.

3.4.1 Double Sideband-Suppressed Carrier Amplitude Modulation

As the name suggests, double sideband-suppressed carrier amplitude modulation (DSB-SC AM) contains two bands of original information and the carrier frequency is suppressed. From the fundamental perspective of frequency content, i.e., frequency spectrum of any signal, there exist two bands of the same information, one at the positive frequency band and the other at the negative frequency band. Using DSB-SC AM, the complete spectrum of baseband signal is translated to carrier frequency and negative carrier frequency as well; however, the carrier frequency component is not present. That is why the technique is known as DSB-SC AM. This modulation scheme is implemented using the time domain multiplication property of Fourier transform. It says that when two signals are multiplied in time domain, the spectra of both the signals are convolved, i.e., the signals are convolved in frequency domain. Mathematically, a DSB-SC AM signal is represented as:

$$S_{DSB}(t) = m(t) A_c \cos(\omega_c t) \tag{3.2}$$

where $m(t)$ is the message/modulating signal, A_c is the amplitude of carrier signal, ω_c is the angular frequency of carrier signal, and t is the time.A sinusoidal message signal and corresponding DSB-SC AM signal is shown in Figure 3.2. The frequency translation of the message signal in this process is shown in Figure 3.3(b).

The frequency translation can be explained by representing the sinusoidal carrier signal using complex exponential signals as

$$\begin{aligned} S_{DSB}(t) &= m(t) A_c \frac{e^{j\omega_c t} + e^{-j\omega_c t}}{2} \\ &= \frac{m(t) A_c}{2} e^{j\omega_c t} + \frac{m(t) A_c}{2} e^{-j\omega_c t} \end{aligned} \tag{3.3}$$

Using the properties of Fourier transforms, it is clearly seen that the first term in the above expression represents the message signal frequency translated to ω_c, and the second term represents the message signal frequency translated to $-\omega_c$. So, using DSB-SC, the frequency of message ω_m signal is translated to $\omega_c + \omega_m$ and $\omega_c - \omega_m$. Here, setting the carrier frequency also has a relation to the bandwidth of the message signal. From the frequency spectra of message signal and the modulated signal as shown in Figure 3.3, it is observed that $\omega_c \geq B$ where B is the bandwidth of the baseband message signal. But, merely setting the carrier frequency satisfying this condition would not be sufficient for the modulated signal to be in the form appropriate for transmission. The carrier frequency, in fact, should be much higher than the

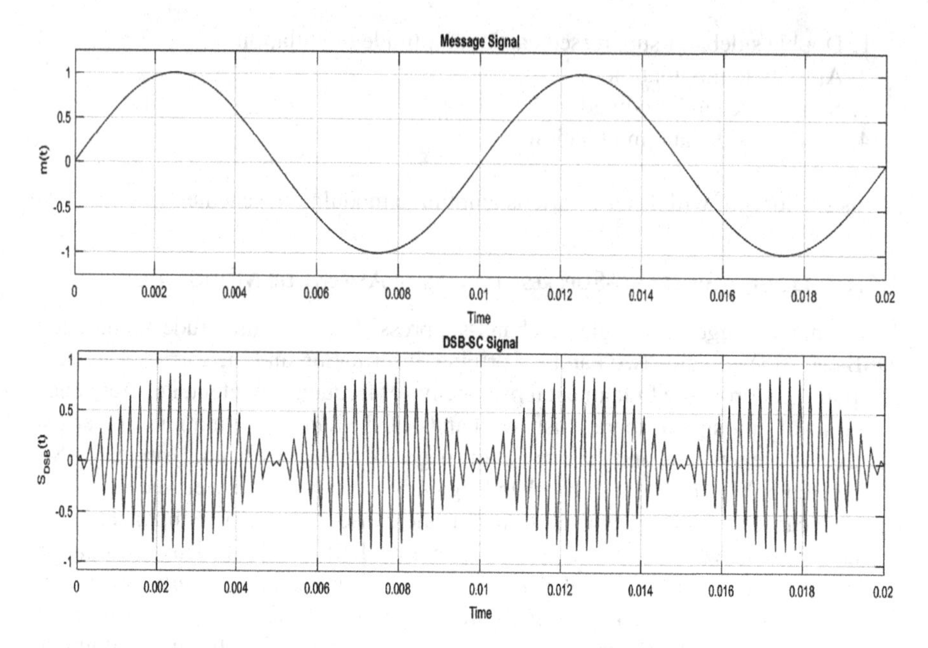

FIGURE 3.2 DSB-SC signals with $m(t) = \sin(\omega_m t)$ and $A_c = 1$.

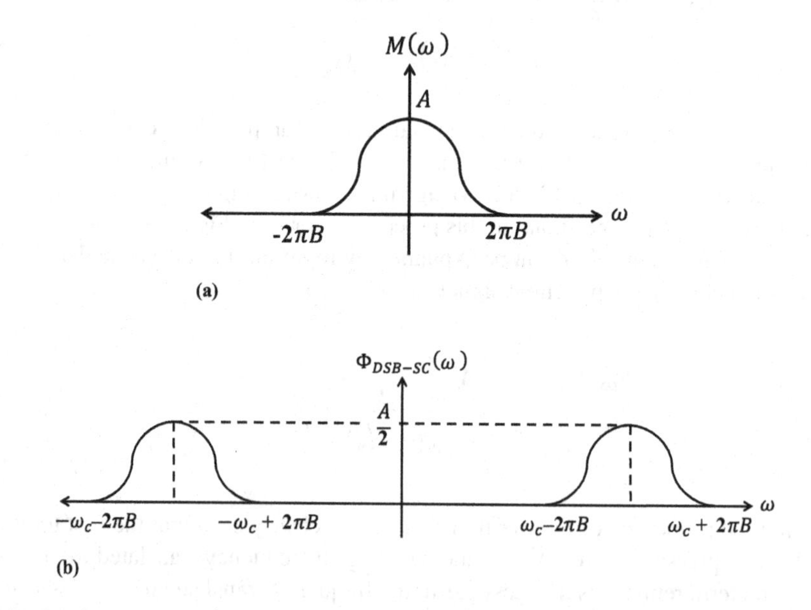

FIGURE 3.3 Frequency spectra of signal before and after DSB-SC modulation: (a) frequency spectrum of message signal, (b) frequency spectrum of DSB-SC signal.

bandwidth of the baseband message signal in order to get a modulated signal in the form which is appropriate for transmission. This is because the required antenna dimensions are proportional to the wavelength of the signal to be transmitted as discussed in Section 3.3.1.2.

3.4.2 DSB-SC Modulators

Analog Multiplier Based Modulators: As per the definition of DSB-SC signals in (3.2), the DSB-SC modulators can be easily realized using an analog multiplier. A block diagram of DSB-SC modulator using an analog multiplier is shown in Figure 3.4. Furthermore, the analog multiplication can be implemented using operational amplifiers (OPAMPs) as well as using integrated chips (ICs) available in market. One of the popular ICs for analog multiplication for generating DSB-SC signals is AD 633 IC [1]. IC AD 633 has various other applications than analog multiplication which we will discuss only if the need arises to explain some functionalities. A schematic for a DSB-SC modulator using IC AD 633 is shown in Figure 3.5. We will also see that the same IC can be used for demodulation of a DSB-SC signal in order to get back the message signal at the receiver. Also, there are other circuit designs which can perform DSB-SC modulation without involving multiplication operation.

Non-linear Devices Based Modulators: Using non-linear devices followed by filters, DSB-SC signals can be generated. We know that diodes, bipolar junction transistors (BJTs), and field-effect transistors (FETs) are non-linear electronic devices. Nowadays, FETs have wide applications in circuit implementations because of their robust and low power operation. Let us take an example of a device which is governed by the following law

$$y(t) = ax(t) + bx^2(t). \tag{3.4}$$

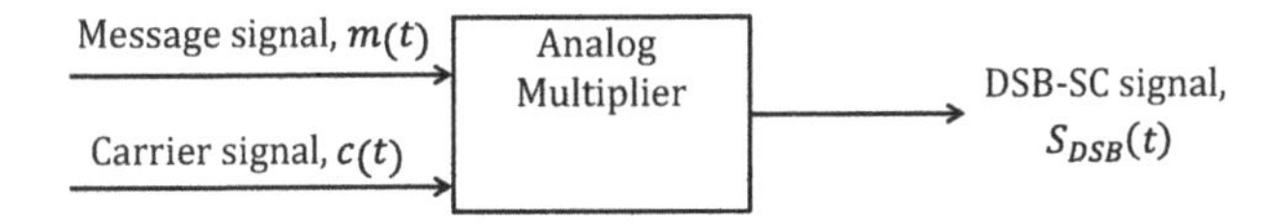

FIGURE 3.4 Block diagram for analog multiplier based DSB-SC modulator.

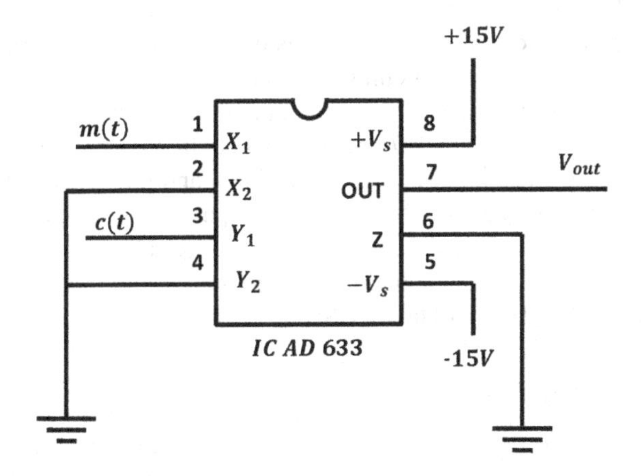

FIGURE 3.5 DSB-SC modulator using IC AD 633.

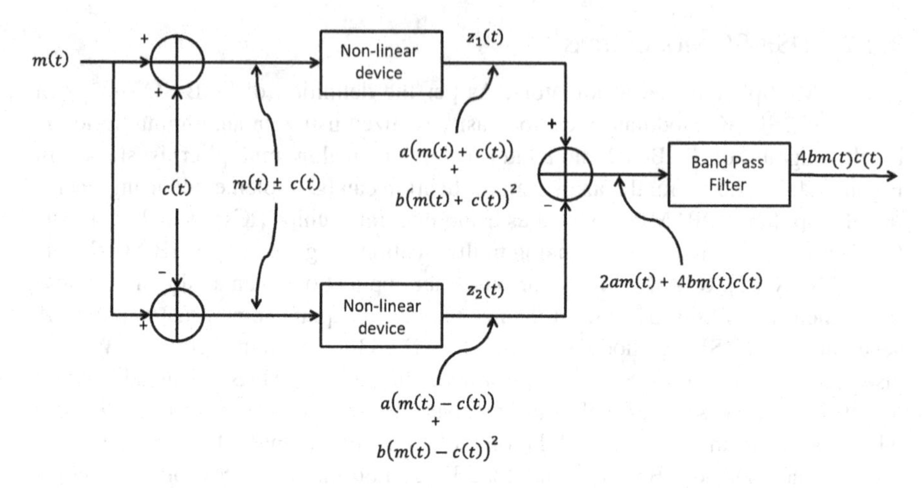

FIGURE 3.6 Block diagram for non-linear devices based DSB-SC modulator.

Using such a device, a DSB-SC signal can be generated with the help of an adder and a bandpass filter (BPF) in addition to the non-linear device as shown in Figure 3.6. In such modulators, the input to the non-linear device is the addition of the message signal, $m(t)$, and the carrier signal, $c(t) = \cos(\omega_c t)$.

So, the output of the linear device is

$$z_1(t) = a(m(t) + c(t)) + b(m(t) + c(t))^2$$

$$z_2(t) = a(m(t) - c(t)) + b(m(t) - c(t))^2 \tag{3.5}$$

Furthermore, the input to the BPF does not contain the carrier signal, $c(t) = \cos(\omega_c t)$; but, it contains two parts, message signal and the DSB-SC signal components. The message signal, being at lower frequency, gets filtered through BPF and the output is $4bm(t)\cos(\omega_c t)$ which is a DSB-SC modulated signal.

Switching Modulators: As the name suggests, this type of modulators uses switching mechanisms of devices like diodes, transistors, etc. to obtain modulated signals. The working principle behind switching modulators is based on the fact that a periodic signal can be represented as a linear combination of sinusoids or complex exponential signals with fundamental frequency and its harmonics known as 'Fourier series representation'. The input is a message signal and it is witched to pass to the output at the carrier frequency. The switched output is equivalent to the message signal multiplied with a pulse train of frequency, ω_c. So, in this method of generation of DSB-SC, a sinusoidal carrier signal is not used. Furthermore, the switched output can be passed through BPF to obtain a DSB-SC signal.

3.4.3 DSB-SC Demodulators

Demodulation is the reverse process of modulation. In modulation, the frequency spectrum of message signal is translated to a higher frequency. While in demodulation, the frequency spectrum is translated to baseband frequency of message signal. So, both processes of modulation and demodulation involve a common process of frequency translation; the only difference is one translates to higher frequency and the other translates to lower frequency.

Like modulation, demodulation can also be realized using multiplier circuits. A DSB-SC modulated signal is given as

$$S_{DSB}(t) = m(t)\cos(\omega_c t) \tag{3.6}$$

when we further multiply this signal with the carrier signal, we get

$$\begin{aligned} D_{DSB}(t) &= m(t)\cos(\omega_c t)\cos(\omega_c t) \\ &= m(t)\left[\frac{e^{j\omega_c t} + e^{-j\omega_c t}}{2}\right]\left[\frac{e^{j\omega_c t} + e^{-j\omega_c t}}{2}\right] \end{aligned} \tag{3.7}$$

The above expression can be further simplified to be represented as

$$\begin{aligned} D_{DSB}(t) &= \frac{m(t)}{4}\left[e^{j2\omega_c t} + e^{-j2\omega_c t} + 2\right] \\ &= \frac{m(t)}{2} + \frac{m(t)}{2}\left[\frac{e^{j2\omega_c t} + e^{-j2\omega_c t}}{2}\right] \\ &= \frac{m(t)}{2} + \frac{m(t)}{2}\cos(2\omega_c t) \end{aligned} \tag{3.8}$$

The above signal is a combination of message signal and a DSB-SC signal modulated with carrier frequency, $2\omega_c$. The second term can be filtered out using a lowpass filter to obtain the message signal part at the output as $\frac{m(t)}{2}$. The block diagram of DSB-SC demodulator is shown in Figure 3.7. Note that in such demodulation systems, a synchronous carrier signal (in both phase and frequency) is required to obtain exact replica of the message signal at the receiver. Such demodulation process is known as 'coherent demodulation'.

Like modulation, DSB-SC signals can also be demodulated using switching circuits.

3.4.4 Amplitude Modulation

Unlike in DSB-SC modulated signals, amplitude modulated signals contain a carrier signal as a part of modulated signals. An AM signal is mathematically represented as

$$\begin{aligned} S_{AM}(t) &= A_c\cos(\omega_c t) + m(t)\cos(\omega_c t) \\ &= (A_c + m(t))\cos(\omega_c t) \end{aligned} \tag{3.9}$$

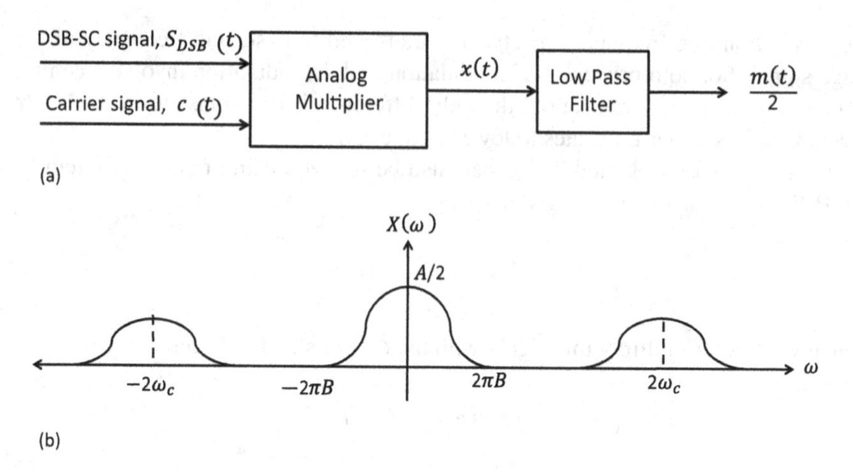

FIGURE 3.7 DSB-SC demodulation: (a) block diagram of DSB-SC demodulator, (b) frequency spectrum of product of DSB-SC signal and carrier signal.

A block diagram of an AM generator using a DSB-SC modulator is shown in Figure 3.8 where the dashed box represents the DSB-SC modulator. The frequency spectrum of an AM signal is basically the same as that of DSB-SC modulated signals except the components at carrier frequency due to the explicit presence of carrier signal component in the modulated signal representation. Another important difference of AM as compared to DSB-SC is that the envelope of the AM signal follows the message signal when $A_c + m(t) > 0$ for all time instants. This enables the receiver to demodulate an AM signal non-coherently. The possibility of non-coherent demodulation simplifies the receiver hardware to a significant amount and hence AM signals have popularity in broadcast applications like radio and TV broadcast. The AM signals with $A_c + m(t) > 0$ for all t and $A_c + m(t) < 0$ for some t are demonstrated in Figure 3.9. From the figure, it is clear that extracting the envelope of the AM signal is enough to recover the message signal from the modulated signal when $|A_c + m(t)| = A_c + m(t)$, i.e. $A_c + m(t) > 0$ for all t, i.e., modulation index less than or equal to 1.

Let us analyze the condition $A_c + m(t) > 0$ for all t in terms of carrier signal and message signal.

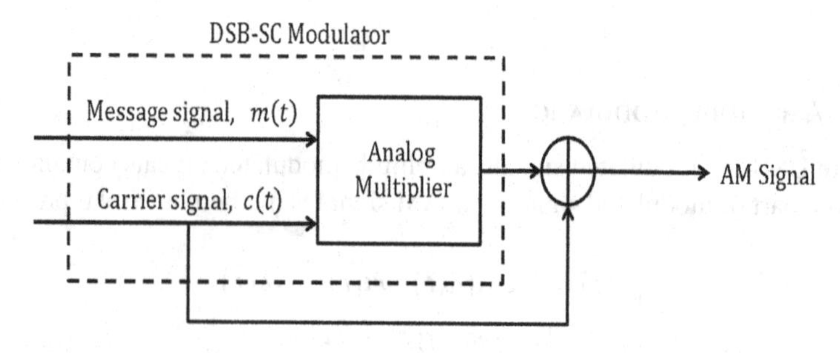

FIGURE 3.8 Block diagram of AM signal generation using DSB-SC modulator.

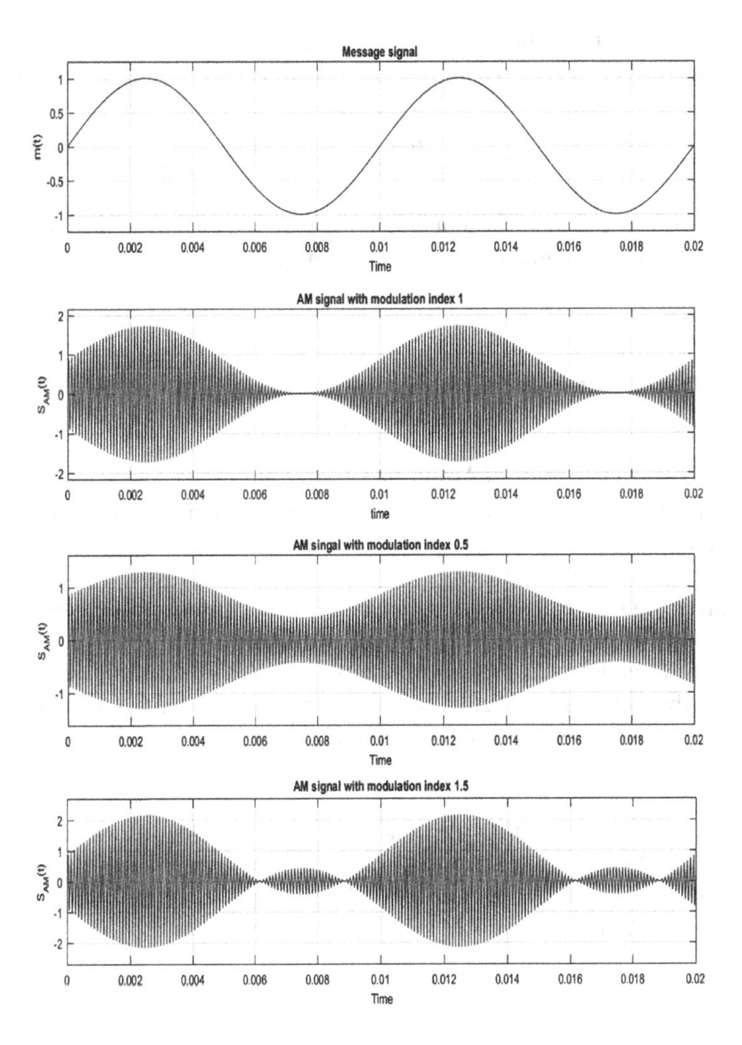

FIGURE 3.9 AM signals with $m(t) = A_m \sin(\omega_m t)$ and $A_c = 1$. The AM signals with modulation index 0.5 and 1.5 represent cases of under-modulation (with $A_m = 0.5$) and over-modulation (with $A_m = 1.5$).

$$A_c + m(t) > 0 \Rightarrow A_c + \min(m(t)) > 0$$

where min $(m(t))$ is the minimum value of message signal.

To satisfy the above expression, we define modulation index as

$$\mu = \frac{\min(m(t))}{A_c} \tag{3.10}$$

For the case that the message signal has zero offset or zero DC value, i.e., $\int m(t)\,dt = 0$ for a given interval which is usually the case for many practical signals, the maximum and minimum values of the message signal are related as $A_m = |\min(m(t))| = \max(m(t))$. In this case, the modulation index is defined as

$$\mu = \frac{A_m}{A_c} \tag{3.11}$$

Again, for non-coherent demodulation, the required condition is $A_c > A_m$. Thus, the similar condition on modulation index can be given as

$$0 < \mu \leq 1$$

where the extreme conditions of the inequality are satisfied as $\mu = 0$ when $A_m = 0$, i.e., message signal is always zero and $\mu = 1$ when $A_m = A_c$.

3.4.4.1 Power Efficiency of Amplitude Modulation

Power efficiency is an important parameter for any transmission technique. In AM, two components are transmitted:

1. Carrier signal which does not contain any useful information given by $A_c \cos(\omega_c t)$.
2. DSB-SC modulated signal which is also known as sideband signal, and it contains the message signal component. This part is responsible for carrying information to be transmitted to the receiver. The sideband signal is given by $m(t)\cos(\omega_c t)$.

It is important to note that for non-coherent demodulation, which is a major advantage of AM, it must satisfy $\mu \leq 1$. The AM signal power has two components: carrier power, P_c, and sideband power, P_m. Sideband power contains useful information. Thus, power efficiency of AM signals, η_{AM}, is defined as

$$\eta_{AM} = \frac{P_m}{P_m + P_c} \tag{3.12}$$

For a special case of tone modulation, i.e., message signal contains single frequency component, e.g., $m(t) = A_m \cos(\omega_m t)$, the efficiency of AM signal is given as

$$\eta_{AM} = \frac{\frac{A_m^2}{4}}{\frac{A_m^2}{4} + \frac{A_c^2}{2}} = \frac{A_m^2}{A_m^2 + 2A_c^2} \tag{3.13}$$

The above expression can be represented in terms of modulation index, μ, as

$$\eta_{AM} = \frac{\mu^2}{\mu^2 + 2} \tag{3.14}$$

From the expression of power efficiency of AM signals considering the tone modulation, it is observed that the efficiency increases with an increase in the modulation index. So, the maximum efficiency is achieved corresponding to the highest possible value of modulation index, $\mu = 1$. Correspondingly, maximum power efficiency is obtained only to be 33%. It turns out that AM transmission is not a power efficient technique, as at least two-thirds of the transmitted power is used in transmission of the carrier signal. But, we will see that this technique brings a significant simplicity in the receiver/demodulator hardware which is making AM transmission of practical importance when it comes to broadcast applications.

3.4.4.2 AM Modulators

AM modulators can be implemented in the same way as that of DSB-SC modulators. The extra circuitry required is to implement the addition of a carrier. A block diagram of AM modulators using DSB-SC modulator is shown in Figure 3.8.

Again, IC AD633 can be used for AM signal generation for the fact that IC AD633 contains an adder which can add one of the inputs to the product of two inputs. So, using IC AD633 for AM signal generation eliminates further need of an adder to add carrier signal to the product of message signal and carrier signal.

3.4.4.3 AM Demodulation

An AM signal is represented as

$$S_{AM}(t) = (A_c + m(t))\cos(\omega_c t). \tag{3.15}$$

We can use the same principle of demodulation as that used for DSB-SC signal demodulation and obtain $\frac{A_c + m(t)}{2}$ by multiplying the AM signal with carrier signal at the receiver and pass the product through a lowpass filter as described in Figure 3.7. Additionally, $\frac{A_c + m(t)}{2}$ can be passed through a DC blocking filter which is a simple capacitor based filter to remove the DC component to obtain $\frac{m(t)}{2}$. But, this coherent way of demodulation uses a carrier signal which is not cost effective for AM signal demodulation.

The AM signals can be demodulated using a non-coherent demodulation technique which eliminates the requirement of carrier signal generation at the receiver. The advantages of non-coherent a demodulation technique are as follows:

- The demodulation hardware is simple.
- It does not require a local oscillator to generate carrier signal at the receiver.
- It does not require synchronization assembly to synchronize the phase and frequency of the locally generated carrier signal at the receiver with the phase and frequency of the received signal.
- The simple hardware makes the receiver cost-effective.

Here, we use the fact that the envelope of the modulated signal in AM follows the message signal exactly. So, if we extract the envelope of an AM signal, we get the

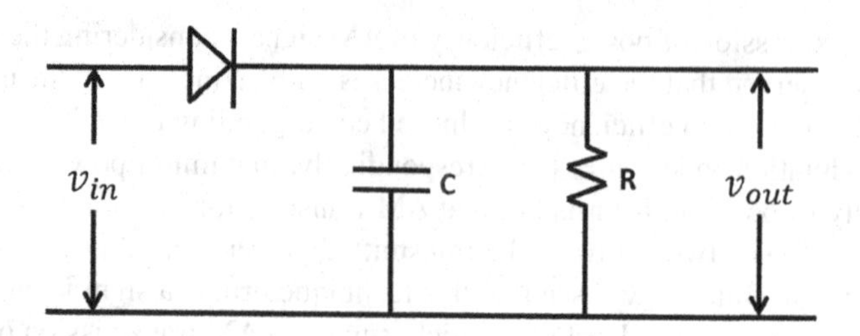

FIGURE 3.10 Envelope detector circuit for AM demodulation.

message signal. This can be done by envelope detection circuit which is a simple combination of diode, capacitor, and resistor as shown in Figure 3.10. The circuit works in the following way:

- Diode functions as a half wave rectifier and allows only positive parts to pass through.
- Capacitor is used to hold the peak value when the diode is reverse biased (off), i.e., when the AM signal is having negative values.
- Resistor gives discharge path to the charge stored on the capacitor during the negative cycle of AM signal. This facilitates the envelope detection even if the peak value of the next positive cycle is less than the peak value of the current positive cycle.

Care should be taken here regarding the *RC* time constant of the envelope detector so that the output of the envelope detector does not contain too many ripples and it follows the message signal, i.e., envelope of AM signal even when the message signal is a decreasing function. So, the trade-off on *RC* time constant can be explained with the following conditioning:

1. Low values of *RC* time constant increase the ripples in the output. Ripples are the high frequency components. To reduce the ripples, the *RC* time constant is to be designed considering the carrier frequency using the relation $RC \gg \frac{1}{\omega_c}$.
2. Large values of *RC* time constant are better to reduce the ripples in the output. However, the output fails to follow the envelope in the duration when the message signal is a decreasing function for too large values of *RC* time constant. To avoid this problem, the *RC* time constant is to be designed considering the message signal bandwidth or the highest frequency component of the message signal, say, $2\pi B$ as $RC < \frac{1}{2\pi B}$ where B is the bandwidth of the message signal expressed in hertz.

Based on the above conditioning, the final designed value of the *RC* time constant for envelope detector should satisfy for proper demodulation of AM signal.

$$\frac{1}{2\pi B} > RC \gg \frac{1}{\omega_c} \tag{3.16}$$

Further, the output of the envelope detector contains the DC offset and some ripples. The DC offset can be blocked using a DC blocking filter which is a simple capacitor based filter. The ripples, as high frequency components, can be further filtered using a lowpass filter.

3.4.5 Single Sideband Modulation

We discussed the two classes of amplitude modulation, DSB-SC and AM. Observing the frequency spectrum of DSB-SC and AM signals, we see that the required bandwidth to transmit both these signals is double the bandwidth of the message signal. In this section, we will introduce single sideband (SSB) modulation which requires half the bandwidth of DSB-SC and AM signals for transmission, i.e., SSB modulation requires the same bandwidth as that of the message signals. So, SSB modulation scheme is also known to be one of the bandwidth efficient modulation schemes for the class of amplitude modulation.

To understand SSB modulation, consider the frequency spectrum of message signal as shown in Figure 3.3(a) also reproduced in Figure 3.11 to maintain the continuity of the readers. This frequency spectrum has both the positive and the negative frequency components. We represent the positive frequency component as $M_+(\omega)$ and the negative frequency component as $M_-(\omega)$ as shown in Figure 3.11. Furthermore, using the positive frequency component and the negative frequency component of the frequency spectrum of the message signal the SSB modulated signal can be represented in the frequency domain as follows:

$$\Phi_{LSB}(\omega) = M_-(\omega - \omega_c) + M_+(\omega + \omega_c)$$

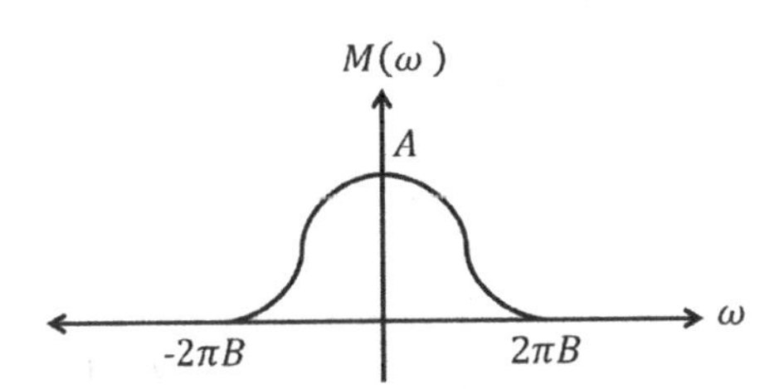

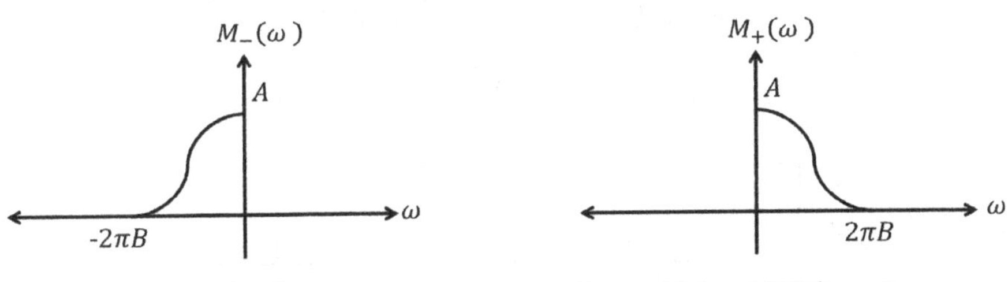

FIGURE 3.11 Frequency spectrum of message signal and corresponding lower sideband (LSB) and upper sideband (USB).

$$\Phi_{USB}(\omega) = M_+(\omega - \omega_c) + M_-(\omega + \omega_c) \quad (3.17)$$

where $\Phi_{LSB}(\omega)$ represents lower sideband (LSB) modulation which can be seen as the part of the spectrum of DSB-SC modulated signals only with the frequencies below ω_c, while $\Phi_{USB}(\omega)$ represents the upper sideband (USB) modulated signal spectrum. Again, the USB signal spectrum can be considered as the part of the DSB-SC spectrum with frequencies above ω_c. The spectrum of SSB modulated signals is shown in Figure 3.12 for LSB modulation, $\Phi_{LSB}(\omega)$, and USB modulation, $\Phi_{USB}(\omega)$.

Ideally, it is possible to obtain SSB signals from DSB-SC signals with the help of filters with sharp/step transition from passband to stop band. But, such filters do not exist in practice; therefore, we use alternative methods to generate SSB modulated signals. For mathematical representation of SSB signals, we further represent $M_+(\omega)$ and $M_-(\omega)$ as

$$M_+(\omega) = M(\omega)u(\omega)$$

$$M_-(\omega) = M(\omega)u(-\omega) \quad (3.18)$$

where $u(\cdot)$ is the unit step function. Also, unit step function can be represented as $u(x) = \frac{1}{2}[1 + sgn(x)]$ and $u(-x) = \frac{1}{2}[1 - sgn(x)]$ where $sgn(\cdot)$ is sign function which has value +1 if the argument is positive signed real number, else it takes value −1.

The purpose of representing the unit step function using sign function is that the transform called 'Hilbert transform' can be used to represent the SSB modulated signals and this gives a practically suitable method for generation of SSB signals. The Hilbert transform of a signal, $x(t)$, is given in the time domain and frequency domain respectively as

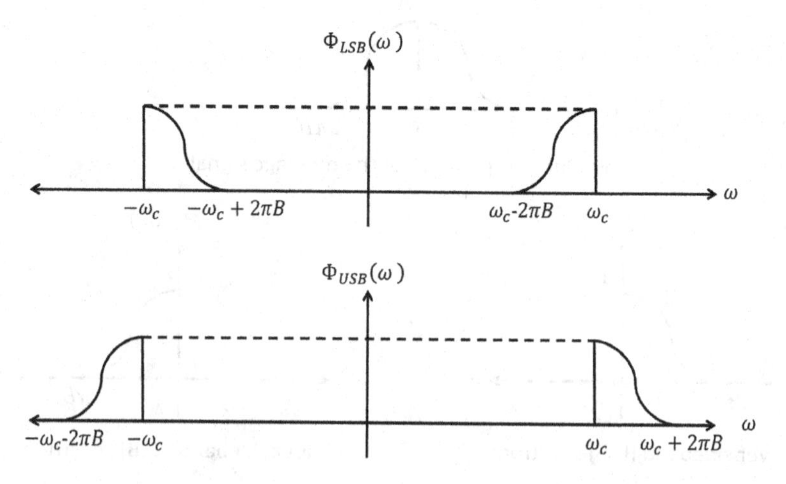

FIGURE 3.12 Frequency spectra of SSB modulated signals for lower sideband (LSB) modulation, $\Phi_{LSB}(\omega)$, and upper sideband (USB) modulation, $\Phi_{USB}(\omega)$.

$$H\{x(t)\} = \frac{1}{\pi}\int_{-\infty}^{-\infty} \frac{x(a)}{t-a}\,da$$

$$X_h(\omega) = -jX(\omega)\,sgn(\omega) \tag{3.19}$$

where $H\{x(t)\}$ is the Hilbert transform of $x(t)$, and $X_h(\omega)$ is the Fourier transform of $H\{x(t)\}$.

The properties of the Hilbert transform can be summarized as follows based on the above definition:

1. The Hilbert transform does not alter the magnitude spectrum of the signal.
2. The Hilbert transform introduces phase shift of $\frac{\pi}{2}$ to the negative frequency components.
3. The Hilbert transform introduces phase shift of $-\frac{\pi}{2}$ to the positive frequency components.

So, SSB signal spectra can be represented in the form of Hilbert transform of the message signal, $M_h(\omega)$, as

$$\Phi_{LSB}(\omega) = \frac{1}{2}\left[M(\omega-\omega_c) - jM_h(\omega-\omega_c)\right] + \frac{1}{2}\left[M(\omega+\omega_c) + jM_h(\omega+\omega_c)\right]$$

$$\Phi_{USB}(\omega) = \frac{1}{2}\left[M(\omega-\omega_c) + jM_h(\omega-\omega_c)\right] + \frac{1}{2}\left[M(\omega-\omega_c) - jM_h(\omega-\omega_c)\right] \tag{3.20}$$

Using the above representation and the definition of Hilbert transform, SSB signals can be represented in the time domain as follows:

$$S_{LSB}(t) = m(t)\cos(\omega_c t) + H\{m(t)\}\sin(\omega_c t)$$

$$S_{USB}(t) = m(t)\cos(\omega_c t) - H\{m(t)\}\sin(\omega_c t) \tag{3.21}$$

Using the phase shifting property of Hilbert transform and the above expressions, LSB and USB modulated signals can be obtained using a combination of phase shifters, analog multipliers, and adders.

SSB Demodulation: SSB signals can be demodulated using a coherent demodulation technique as discussed for demodulation of DSB-SC signals. This includes multiplying the SSB signal with the carrier signal and passing the product through a lowpass filter to remove the high frequency components.

3.5 QUADRATURE AMPLITUDE MODULATION

Quadrature amplitude modulation (QAM) is another method to improve the bandwidth efficiency of an amplitude modulated signal. QAM is an alternative to SSB for its simplicity in generation. QAM uses the fact that if two carriers with a quadrature phase are used to modulate two different message signals, both message signals can be separated without an interference. In QAM, sine and cosine are used as the two quadrature phased carriers to modulate two message signals, say, $m_1(t)$ and $m_2(t)$ using a DSB-SC modulator. A QAM signal can be represented as

$$S_{QAM}(t) = m_1(t)\cos(\omega_c t) + m_2(t)\sin(\omega_c t) \tag{3.22}$$

So, using QAM, we get the same bandwidth efficiency as that of SSB modulation, i.e., two message signals are transmitted using the same carrier frequency in case of QAM.

Additionally, QAM signals are demodulated coherently by multiplying $S_{QAM}(t)$ with the respective carrier signal and passing the product through lowpass filter to obtain the corresponding message signal. The modulation and demodulation process for QAM needs the following conditions for exact extraction of $m_1(t)$ and $m_2(t)$:

1. The sine and cosine carriers used for modulation must be synchronous to each other. Any delay or advance in terms of time and/or phase results in interference between $m_1(t)$ and $m_2(t)$.
2. At the demodulator, the sine and cosine carriers must synchronize to each other as well as with the received signal phase and frequency for exact demodulation of message signals without interference.

3.6 FREQUENCY DIVISION MULTIPLEXING

Multiplexing is a technique to transmit multiple signals using the same connection/line. In case of wireless communication, the medium between transmitter and receiver is known as 'channel'. So, multiplexing is referred to as the transmission of multiple signals of similar nature over a channel. In frequency division multiplexing (FDM), a band of frequency is divided into channels. Each channel is available to modulate different message signals. Also, the modulated messages at different channels are transmitted simultaneously. The separation between two channels is kept such that the spectrum of modulated signals for adjacent channels do not overlap to avoid any interference between signals transmitted at the adjacent channel frequencies. A demonstration of multiplexing three signals with a block diagram of FDM system is shown in Figure 3.13. A frequency spectrum of FDM of three signals $x_1(t)$, $x_2(t)$, and $x_3(t)$ respectively modulated at carrier frequencies f_{c1}, f_{c2}, and f_{c3} is shown in Figure 3.13(b).

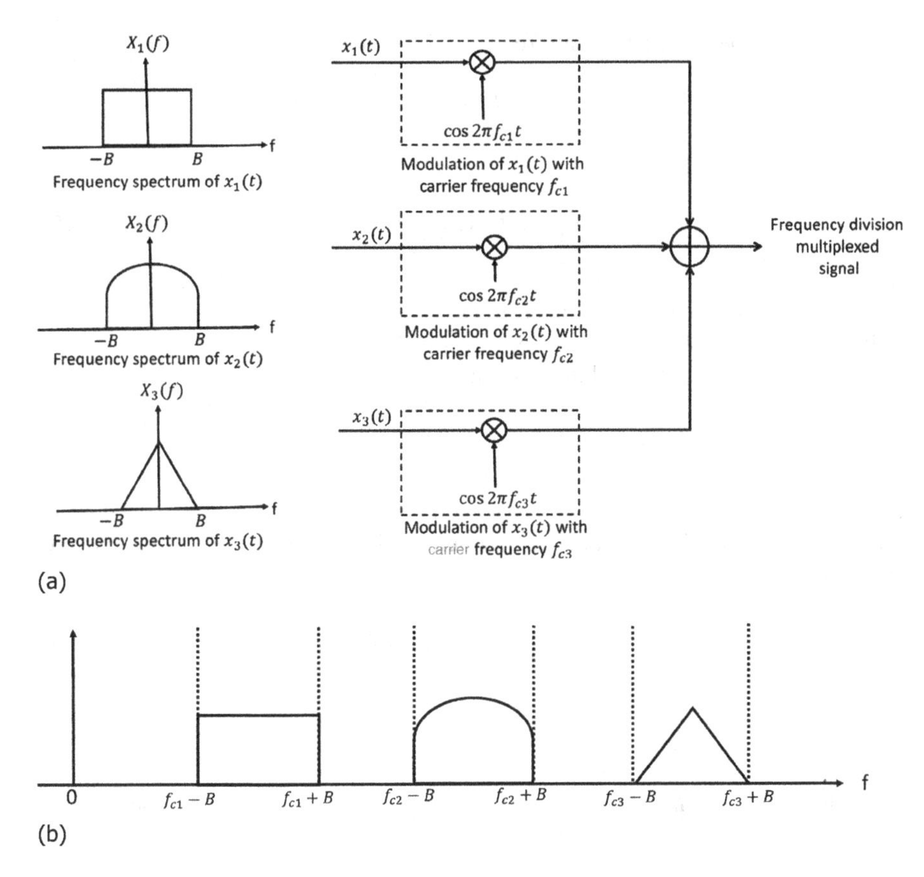

FIGURE 3.13. Frequency division multiplexing: (a) demonstration of frequency division multiplexing (FDM) of three messages modulated at frequencies f_{c1}, f_{c2}, and f_{c3}, (b) frequency spectrum of FDM signal.

3.7 ANGLE MODULATION

If we represent the carrier signal as

$$c(t) = A_c \cos(\omega_c t + \varphi), \tag{3.23}$$

the angle component has two parameters, the frequency, ω_c, and the phase, φ. So, the modulations of this type are classified as frequency modulation (FM) or phase modulation (PM) based on the parameter which is varied according to the instantaneous value of the message signal.

For the above expression, the instantaneous angle of the carrier signal is given as

$$\theta_i(t) = \omega_c t + \varphi \tag{3.24}$$

and the relation between angle and frequency is given as

$$\omega_i(t) = \frac{d\theta(t)}{dt}$$

$$\theta_i(t) = \int_{-\infty}^{-t} \omega_i(t)dt \tag{3.25}$$

where ω_i is the instantaneous frequency of the signal.

In FM, the instantaneous frequency varies with message signal, i.e., the instantaneous frequency is represented in the form

$$\omega_i(t) = \omega_c + fm(t) \tag{3.26}$$

where f is the constant to control variation in instantaneous frequency. Using the above expression of instantaneous frequency, the instantaneous angle can be given by

$$\theta_i(t) = \omega_c t + f\int_{-\infty}^{-t} m(t)dt. \tag{3.27}$$

3.7.1 Frequency Modulation

Using the discussion and representations in the previous section, a frequency modulated (FM) signal can be represented as

$$S_{FM}(t) = A_c \cos\left(\omega_c t + f\,{}_{-\infty}^{-t}\, m(t)dt\right), \tag{3.28}$$

The instantaneous frequency of the FM signal, $S_{FM}(t)$, can be given as mentioned in expression (3.26). So, the maximum and minimum frequency of a FM signal can be given as

$$\omega_{FM_{min}} = \omega_c + f\min\{m(t)\}$$

$$\omega_{FM_{max}} = \omega_c + f\max\{m(t)\} \tag{3.29}$$

Looking at the first instance to the maximum and minimum frequencies of a FM signal, it is inferred that the bandwidth of a FM signal can be given as $f\max\{m(t)\} - f\min\{m(t)\}$. If the message signal is with zero offset or has zero average value and assuming $\max\{m(t)\} = |\min\{m(t)\}| = M_{max}$, the bandwidth of a FM signal can be given as $2fM_{max}$. But, this is not true. The actual bandwidth of FM signals is much higher than this expected bandwidth based on the difference of maximum and minimum frequencies. Let us discuss the mathematical representation of

PM and an equivalence between FM and PM before going into the details of bandwidth considerations.

3.7.2 Phase Modulation

In phase modulation (PM), the instantaneous phase varies with message signal, i.e., the instantaneous phase is represented in the form

$$\varphi_i(t) = pm(t) \tag{3.30}$$

where p is the constant to control variation in the instantaneous phase. Using the above expression of instantaneous phase, the instantaneous angle and instantaneous frequency can be given by

$$\theta_i(t) = \omega_c t + pm(t)$$

$$\omega_i(t) = \omega_c + p\frac{dm(t)}{dt} \tag{3.31}$$

So, the PM signal can be mathematically represented as

$$S_{PM}(t) = A_c \cos(\omega_c t + pm(t)). \tag{3.32}$$

FM and PM signals are demonstrated in Figure 3.14. FM signal is shown for $f = 1$, and PM signal is shown for $p = 4$ in the figure.

3.7.3 Equivalence between Frequency and Phase Modulation

Comparing the expressions of FM and PM signals as in (3.28) and (3.32), we can develop the following equivalence between frequency and phase modulation:

- The angle of the carrier signal is given as $\omega_c t + f\int_{-\infty}^{t} m(t)dt$ and $\omega_c t + pm(t)$ for FM and PM signals, respectively.
- If the input to the phase modulator is integral of the message signal, we get FM signal using a phase modulator.
- If the input to the frequency modulator is differentiation of the message signal, we get PM signal using a frequency modulator.

So, frequency modulator and phase modulator can be used to generate both the types of angle modulated signals accordingly using the equivalence. In the same manner, designing a demodulator for FM signal can be used to demodulate a PM signal using similar transformations on the demodulated signal to get back the original signal. Considering this equivalence, it is sufficient to analyze either FM signals or PM signals for their bandwidth requirements. A block diagram representation of equivalence between FM and PM is shown in Figure 3.15 to illustrate FM generation

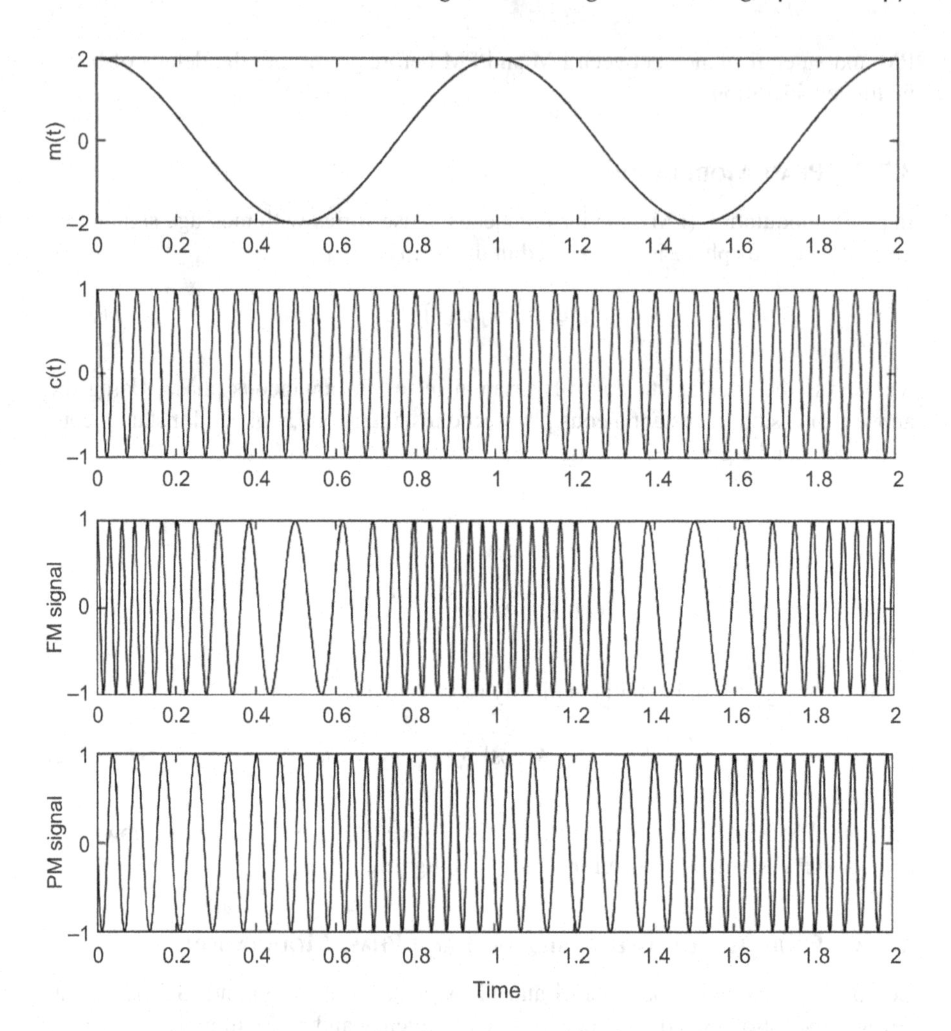

FIGURE 3.14 Frequency modulated and phase modulated signals for sinusoidal message signal, $m(t) = 2\cos(2\pi t)$ and carrier frequency 20 times the message signal frequency.

using a phase modulator in Figure 3.15(a), and PM generation using a frequency modulator in Figure 3.15(b).

3.7.4 Bandwidth of Angle Modulated Signals

We have already established an equivalence between FM and PM. So, we will consider FM signals for the discussions on bandwidth analysis. Similar discussions on the bandwidth analysis of PM signals is left to the readers.

We consider FM signals for the bandwidth analysis of angle modulated signals. Let us represent the FM signal in the form of complex exponential as

$$S_{FM}(t) = A_c \Re e\left\{(e^{j\omega_c t + f}{}_{-\infty}^{\;t} m(t)dt\right\} \tag{3.33}$$

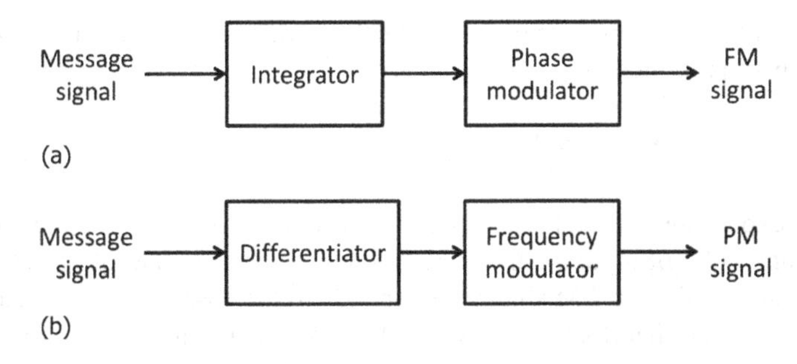

FIGURE 3.15 Block diagram of frequency modulator and phase modulator using equivalence between frequency and phase modulation: (a) FM generation using a phase modulator, (b) PM generation using a frequency modulator.

Let us denote $a(t)=\int_{-\infty}^{t} m(t)dt$ for a compact representation of modulated signal as

$$S_{FM}(t)=A_c\Re e\left\{e^{j(\omega_c t+f\alpha(t))}\right\} \tag{3.34}$$

Expanding the second term of the complex exponential using power series, the above expression can be written as

$$S_{FM}(t)=A_c\Re e\left\{1+jfa(t)-\frac{f^2a^2(t)}{2!}+\cdots+j^n\frac{f^na^n(t)}{n!}+\cdots\left[e^{j(\omega_c t)}\right]\right\} \tag{3.35}$$

Thus, the FM signal can be expressed in the form of an infinite series expansion as

$$S_{FM}(t)=A_c\cos(\omega_c t)-fa(t)\sin(\omega_c t)-\frac{f^2a^2(t)}{2!}\cos(\omega_c t)+\cdots \tag{3.36}$$

We observe from the above expression that the FM signal is represented as a summation of infinite number of DSB-SC signals where each DSB-SC signal is obtained by modulating integer powers of $\alpha(t)$.

Considering the bandwidth of the message signal as B, the bandwidth of $\alpha(t)$ is also B. So, the bandwidth of $\alpha^n(t)$ is nB. Thus, the bandwidth of FM signal is infinite. But, recall the fact that practical signals with infinite bandwidth cannot be transmitted without distortion due to channels with limited bandwidth. So, to consider practical bandwidth of FM signal to be finite, consider that $f\alpha(t)$ has some finite value, the denominator of $\frac{f^n\alpha^n(t)}{n!}$ would be very large, and the contribution to the energy would be very small. So, the signals can be considered to have limited bandwidth.

Another aspect to this gives rise to the concept of narrowband frequency modulation (NBFM). To study NBFM, consider the message signal such that

$$\left|fa(t)\right|\ll 1. \tag{3.37}$$

So, all the terms containing exponents of $fa(t)$ can be neglected and the modulated signal can be represented as

$$S_{FM}(t) \approx A_c\left[\cos(\omega_c t) - fa(t)\sin(\omega_c t)\right] \quad (3.38)$$

The above expression represents a signal similar to the AM signal. So, the minimum possible bandwidth of an angle modulated signal is equal to that of an AM signal. The minimum bandwidth FM signal corresponds to the case where message signal value is very small; therefore, this case has little practical importance. We will now discuss a more accurate bandwidth analysis for practical FM signals.

3.7.4.1 Bandwidth Analysis of Wideband FM Signals

In this section, we consider a generalized case without any restrictions on frequency deviation of modulated signal from the carrier frequency. So, we get rid of the required condition on message signals as discussed earlier, i.e., $|fa(t)| \ll 1$. We usually refer to such FM signals as 'wideband FM signals'.

For this analysis, consider a message signal of bandwidth B Hz. The message signal is approximated as a staircase signal with each stair/cell of duration $\frac{1}{2B}$ seconds. This stair duration is decided in accordance to meet the conditions required for sampling which will be discussed in the next chapter. With this duration of cell, we can pass the staircase approximated signal through a lowpass filter to obtain the original signal back from the approximated signal. Now, consider a cell at an instant t_n. Let the value of the signal at this cell be $\cos\{\omega_c t + fm(t_n)\}$. The spectrum of this pulse at the cell at instant t_n can be given as

$$\frac{1}{2}\text{sinc}\left\{\frac{\omega + \omega_c + fm(t_n)}{4B}\right\} + \frac{1}{2}\text{sinc}\left\{\frac{\omega - \omega_c fm(t_n)}{4B}\right\} \quad (3.39)$$

where sinc$(\cdot)$ is sinc function defined as $\text{sinc}(x) = \frac{\sin x}{x}$.

The bandwidth of FM signals can be analyzed based on the above spectrum. Note that the frequency domain representation of every stair/cell has the same bandwidth as that of a rectangular pulse of duration $\frac{1}{2B}$ seconds. So, the bandwidth of each stair/cell would be $4B$ Hz. But in FM signals with staircase approximated message signal, the location of spectrum of each stair/cell is different, and hence the actual bandwidth of the FM signal is more than the bandwidth of a rectangular pulse of duration $\frac{1}{2B}$ seconds. For example, the location of the spectrum is $\omega_c + fm(t_n)$ for the cell at an instant t_n.

Let us analyze the locations of spectrum of cells in an FM signal obtained after staircase approximation of message signal. Consider that the message signal has maximum and minimum values of m_{max} and m_{min}, respectively. So, the extreme locations of sinc functions corresponding to the spectra of cells with maximum value and minimum value of messages are $\omega_c + fm_{max}$ and $\omega_c + fm_{min}$, respectively. So the bandwidth of FM signals can be given as

$$B_{FM} = 4B + \frac{f\left(m_{max} - m_{min}\right)}{2\pi} \tag{3.40}$$

For the NBFM case, we have already shown that the bandwidth is $2B$ Hz. But, looking at the above expression, we get the minimum bandwidth of $4B$ Hz when the second term is zero, i.e., at its minimum value. So, it is apparent that due to staircase estimation of message signal, we get extra bandwidth in our analysis. So, to further obtain an accurate bandwidth estimation of FM signals, we use the correction factor so as to match the NBFM bandwidth that we analyzed earlier. An accurate expression for bandwidth of signals can be given as

$$B_{FM} = 2B + \frac{f\left(m_{max} - m_{min}\right)}{2\pi} \tag{3.41}$$

The above expression is widely known as 'Carson's rule' and also is represented as $B_{FM} = 2(B + \delta f)$ where Δf is frequency deviation defined as

$$\Delta f = \frac{f\left(m_{max} - m_{min}\right)}{4\pi}$$

Based on the frequency deviation, deviation ratio is defined as $\beta = \frac{\Delta f}{B}$ which is also the relative frequency deviation. For tone modulation, deviation ratio is referred to as 'modulation index' for FM signals.

3.7.5 FM Generation

FM generation can be done using (1) a direct method and (2) an indirect method. We will discuss each of these methods next.

3.7.5.1 Direct Method for FM Generation

In direct method of FM generation, we use voltage controlled oscillators (VCOs). Oscillators are usually designed to generate signals with particular frequency. VCOs are the type of oscillators that have an input to control the frequency of the signal that is generated. The purpose of FM is also the same, i.e., to change the frequency of the carrier signal according to the instantaneous value of the message signal. So, for FM generation using a VCO, a message signal can be given at the input of a VCO and FM signal can be obtained at the output.

3.7.5.2 Indirect Method for FM Generation

In indirect method of FM generation, we first generate an NBFM signal. From our analysis of FM signal bandwidth, we know that the NBFM signals can be generated in the same way as that of generation of DSB-SC signals. However, the NBFM signals differ from DSB-SC signals in that the amplitude of FM signals is the same all the time. To meet this uniform amplitude requirement, the only extra component required is amplitude limiter so as to keep the amplitude of modulated signal uniform for all the time.

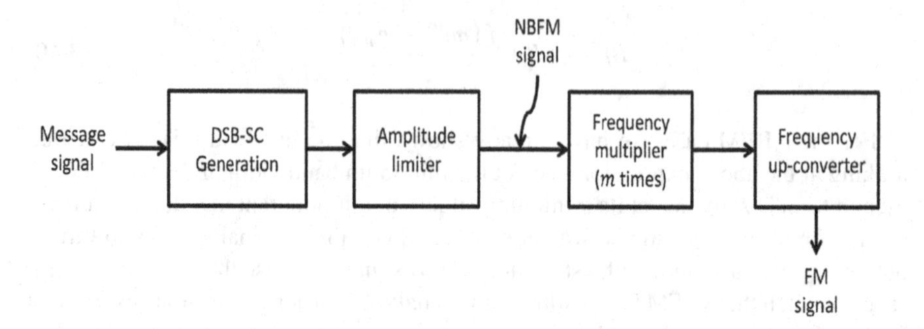

FIGURE 3.16 Block diagram of indirect FM generation using DSB-SC modulator and frequency multiplier(s) and up-converter(s).

NBFM signals have bandwidth that is twice that of message signal bandwidth. Further, NBFM signals are passed through a chain of frequency multipliers and frequency converters to obtain FM signal with required bandwidth and required carrier frequency. Block diagram of indirect FM generation using DSB-SC modulator and frequency multiplier(s) and up-converter(s) is shown in Figure 3.16 where the frequency multiplier block may consists of series cascade of multiple frequency multipliers to obtain total multiplying factor of m (to get m times the frequency of input signal). Thus, it increases the bandwidth of amplitude limited DSB-SC signal by m times. The m is chosen based on the required FM bandwidth. Further, the frequency up-converters may also be multiple in number based on the required carrier frequency. The frequency up-converters may also be inserted in series cascade of frequency multipliers to have multiple up-conversions with multiple bandwidth increments.

3.7.6 FM Demodulation

Let us revisit the mathematical representation of a frequency modulated signal,

$$S_{FM}(t) = A_c \cos\left(\omega_c t + f\int_{-\infty}^{t} m(t)dt\right). \tag{3.42}$$

To extract the information which is embedded in the instantaneous frequency of the signal, we differentiate the FM signal to obtain

$$S'_{FM}(t) = -A_c\left(\omega_c + fm(t)\right)\sin\left(\omega_c t + f\int_{-\infty}^{t} m(t)dt\right). \tag{3.43}$$

The above expression represents an AM plus FM signal with the envelope containing information about the message signal. Note that since the carrier frequency is larger than the message part, i.e., $\omega_c > fm\,(t)$, for all the time instants, the message

At this point of discussion about the angle modulation systems, we note the following:

1. The information of message is contained in the angle part of the carrier signal.
2. The amplitude of the modulated signal (for both FM and PM) is uniform (constant) for all the time, irrespective of the instantaneous value of message signal. So, this kind of modulation scheme is known as 'constant power modulation'.
3. The message signal power affects the bandwidth of the modulated signal.
4. Demodulation of angle modulated signal can be done using a differentiator followed by an envelope detector.

signal can be obtained using envelope detection circuit used in the AM demodulation.

3.8 SUPERHETERODYNE RECEIVERS

The function of receiver is to obtain an accurate copy of the message signal from the received signal. The first block/component of any radio receiver is an antenna. The simplest receiver antenna is a wire antenna, i.e., any conducting wire of suitable length can be used as a receiver antenna. This is the reason in mobile phones with FM radio one needs to connect headphones to start listening to FM channels. Discussion on various kinds of antennas based on application (transmission or reception)/frequency of operation, etc. is out of the scope of this book. Antenna serves as the transducer to convert the signals in the form of electromagnetic waves into electrical form. Further signal processing of these electrical signals is possible at various stages of the receiver.

There are two types of receivers used in communication systems: (1) homodyne and (2) heterodyne.

In homodyne receivers, the RF signal is directly converted to baseband signal. So, such receivers are also referred to as 'single stage receivers'. In such receivers, a local oscillator operates at the carrier frequency in case of coherent detection, while the demodulator circuitry is tuned to the carrier frequency in case of non-coherent demodulation. In heterodyne receivers, there are two stages of frequency conversion. After the first stage, the output is still a modulated signal at an intermediate frequency (IF). The IF signal is further down-converted to obtain baseband signal or the demodulated signal. Such receivers in AM/FM broadcast systems are known as 'superheterodyne receivers'. A block diagram of a superheterodyne receiver is shown in Figure 3.17.

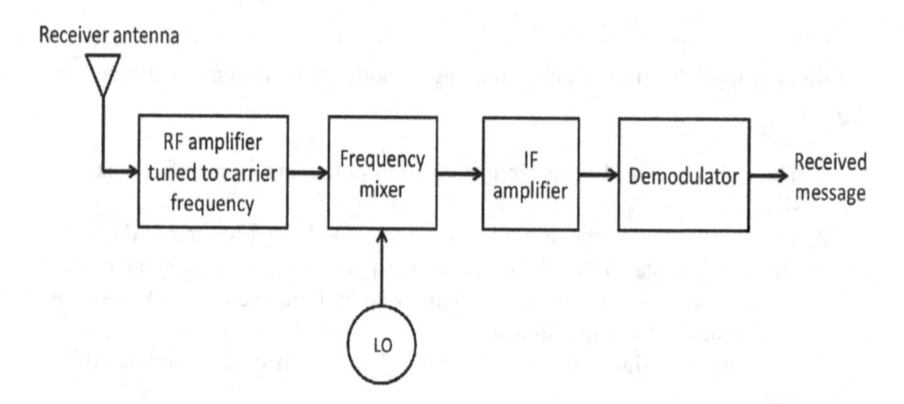

FIGURE 3.17 Block diagram of superheterodyne receiver.

Superheterodyne receivers basically consist of two sections: (1) RF section which operates at the radio frequency (RF) band, i.e., the carrier frequency band, and (2) IF section which operates at the intermediate frequency (IF).

The components of an RF section are

1. RF amplifier,
2. Local oscillator, and
3. Frequency mixer.

The components of an IF section are

1. IF amplifier and
2. Demodulator.

Two stages of frequency down-conversion are helpful to improve the selectivity of the receiver. The antenna receives the intended signal and the other signals present in that particular frequency band, while the receiver has to select only the channel to which it has been tuned. In order to select the required channel, the signal processing and hardware complexity at radio frequency band becomes complex. So, selectivity is also realized at two stages of frequency down-conversion: one at RF amplifier and the other at IF amplifier. Although the selectivity of RF section is very poor, it plays an important role in image frequency rejection through the selection at RF amplifier. Before we discuss the concept of image frequency rejection, we will discuss the operation and purpose of each of the components of RF and IF sections.

- RF amplifier: The input to the RF amplifier comes from the antenna and contains all the frequency components that is received by the antenna. This includes the complete band of frequencies, e.g., an FM receiver antenna receives all the frequencies in the range 88 to 108 MHz while the bandwidth of a single channel for an FM radio is 200 kHz. So, the RF amplifier not only amplifies the signal received by the antenna, but it also tunes its maximum amplification at the carrier frequency of interest. This eliminates some of the nearby channels and reduces the bandwidth of the signal at the output. More specifically, it takes care of the image frequency component.

- Local oscillator: Local oscillator (LO) generates the RF signal with frequency so that the difference of LO frequency and carrier frequency is the same as that of IF. In an FM radio system, IF is set to 10.7 MHz. So, the LO for an FM receiver is tuned to generate frequency equal to f_c +10.7 MHz. Tuning of the respective frequencies of LO and RF amplifier can be paired and is usually realized by changing the capacitance values of the tunable circuits.
- Frequency mixer: The frequency mixer takes two inputs: the signal generated by LO and the amplified RF signal from the RF amplifier. Frequency mixer down-converts the carrier frequency to IF. Mixer is a non-linear operator to produce sum and difference of the input frequencies and their harmonics.
- IF amplifier: Apart from amplification, the most important functionality of the IF amplifier is to select a single channel. The bandwidth of IF amplifier is chosen such that it amplifies the signal belonging to a single channel. For FM radio systems, the IF amplifier bandwidth shall be set to 200 kHz.
- Demodulator: The output of IF amplifier is still a modulated signal. The demodulator converts the IF amplifier output to baseband signal or the message signal depending on a coherent or non-coherent system. Further, the output of the demodulator is used as an input to suitable transducer, example speaker for audio signals and television screen for video signals. The output of the demodulator may be amplified based on requirement of the application or the user. This amplification serves as the volume control for audio signals.

3.8.1 Concept of Image Frequency

Let the carrier frequency of interest be f_c from the output of RF amplifier. To obtain proper IF in this case, it has already been discussed that the LO generates signal with frequency equal to $f_{LO} = f_c + f_{IF}$ where f_{IF} is the IF and f_{LO} is the LO frequency. Now, in such a situation, we get f_{IF} at the mixer output in the following ways:

$$f_{IF} = \left(f_c + f_{IF}\right) - f_c \tag{3.44}$$

$$f_{IF} = \left(f_c + 2f_{IF}\right) - \left(f_c + f_{IF}\right) \tag{3.45}$$

This means that $f_c + 2f_{IF}$ is the other frequency apart from f_c which produces IF at the output of the mixer when we set $f_{LO} = f_c + f_{IF}$. The frequency, $f_c + 2f_{IF}$, is known as 'image frequency'. Image frequency is $2f_{IF}$ apart from the frequency of interest; therefore, it is easy to reject it although little selectivity is available at the RF amplifier or the RF section of the superheterodyne receiver.

3.9 SUMMARY

In this chapter, we note the following points:

- DSB-SC AM is simple frequency translation of message signal spectra using high frequency carrier.
- DSB-SC AM can be demodulated coherently, which makes the receiver design complex and costly.
- AM can be generated by adding carrier signal to DSB-SC signal. AM signals can be demodulated non-coherently. This makes the receiver hardware simple and cost effective.
- AM signal is highly energy inefficient because transmitting the carrier needs at least 67% of power, i.e., the receiver cost effectiveness comes at the cost of power efficiency at the transmitter.
- To improve the bandwidth efficiency, SSB and QAM have been effective candidates.
- SSB and QAM cannot be demodulated non-coherently; therefore, the bandwidth efficiency is improved at the cost of receiver hardware complexity and cost in SSB and QAM.
- Angle modulation schemes are constant amplitude schemes, i.e., the modulated signal is of uniform amplitude all the time.
- Angle modulated signal can be demodulated non-coherently by using a differentiator and an envelope detector. This makes the receiver design using simple hardware.

REFERENCE

1. "Data sheet of IC AD 633," https://www.analog.com/media/en/technical-documentation/data-sheets/AD633.pdf (accessed February 2 2021).

4 Sampling and Analog to Digital Conversion

4.1 INTRODUCTION

We already discussed that all natural signals are analog in nature, while most computer systems are digital. So, there is a need to convert analog signals into digital signals for further processing with better efficiency and effectiveness. Also, this reduces the complexity of integration of various systems.

In this chapter, we will begin with the introduction to digital systems and their advantages. Further, we will see the basic differences between analog and digital communication systems. When a signal is converted to digital form from its natural analog form, some information is lost. However, there are techniques through which the original information can be regenerated through the process of reconstruction.

4.2 ADVANTAGES OF DIGITAL SYSTEMS

The advantages of digital systems are listed as follows:

- Digital storage is easy and less expensive.
- Processing the signals in digital domain is easy. It is done based on simple operations of multiplications and additions.
- Power consumption of digital systems is less.
- Digital systems and signals are more immune to noise.
- Multiplexing of digital signals is comparatively easy compared to analog signals.
- Availability of regenerative repeaters exists in digital systems.
- Encoding in digital signals for error detection, error correction, and data privacy/secrecy is possible.
- Digital systems are reliable.

4.3 DIFFERENCES BETWEEN ANALOG COMMUNICATION SYSTEM AND DIGITAL COMMUNICATION SYSTEM

We already presented the modulated signals for analog communication systems in Chapter 3 as

$$S_{DSB}(t) = m(t) A_c \cos(\omega_c t) \tag{4.1}$$

for double sideband-suppressed carrier (DSB-SC) modulation scheme and

DOI: 10.1201/9781003213468-4

$$S_{FM}(t) = A_c \cos\left(\omega_c t + f\int_{-\infty}^{t} m(t)dt\right), \tag{4.2}$$

for frequency modulation (FM) scheme.

So, suppose we use the modulating signal in digital form; consider binary for a while. There would be only two levels in the message signal, viz., $-A$ volts to represent bit '0' and $+A$ volts to represent bit '1'. Now, considering the above representations of the modulated signals results in only two amplitude levels in the DSB-SC modulated signals and only two frequency levels in FM signals. This shows that the modulated signal would have discrete levels if the message signal is in the digital form.

This also makes it clear that the main difference between digital communication systems and analog communication systems is the form of message signals.

- Message signals are analog in analog communication systems.
- Message signals are digital in digital communication systems.
- Modulated signals can be represented using the same expressions in both analog communication systems and digital communication systems.

4.4 CONVERSION FROM ANALOG TO DIGITAL

Conversion of an analog signal into a digital signal can be done by following the sequence of operations given below:

1. Sampling
2. Quantization
3. Encoding

The whole process of analog to digital conversion is also referred to commonly as pulse code modulation (PCM).

4.4.1 Sampling and Reconstruction

Sampling is the process of converting a continuous time signal into a discrete time signal. There are certain parameters associated with the process of sampling:

1. Sampling interval: The interval at which instantaneous values of the continuous time signal are retained.
2. Sampling frequency: The frequency at which the samples are taken. Sampling frequency is also referred to as 'sampling rate'. Sampling rate is the inverse of sampling interval.
3. Nyquist rate: The minimum sampling rate that enables proper reconstruction of the sampled signal to get back the original copy of continuous time signal.

Reconstruction is the process to get back the original continuous time signal from the sampled signal. It is crucial to select proper sampling frequency while sampling the continuous time signal in order to reconstruct exact copy of the original signal.

4.4.2 Types of Sampling

In this section, we will discuss various types of sampling which include: ideal sampling, natural sampling, and flat top sampling.

4.4.2.1 Ideal Sampling

As the name suggests, ideal sampling is for understanding the process of sampling, and it does not occur practically. In ideal sampling, only the instantaneous values of the continuous time signals at the sampling instant are retained in the sampled signals. This is not possible due to practical limitations of the switching devices. Practical switching does not happen instantaneously, and, hence, it makes retaining only the instantaneous values almost impossible. An example of ideal sampling is shown in Figure 4.1(a).

4.4.2.2 Natural Sampling

In natural sampling, we use a process similar to multiplication of the continuous time signal with pulse train. The pulses are of narrow width and the frequency of arrival of pulses is the same as the sampling frequency. So, when the pulse exists, the sampled signal takes the same values of the continuous time signal. The pulse widths are set so that the whole process takes care of the switching times of the devices used in the sampling process. An example of natural sampling is shown in Figure 4.1(b).

4.4.2.3 Flat Top Sampling

Flat top sampling is a sampling process in which the instantaneous value of the continuous time signals at the sampling instant is retained for a while. Flat top sampling also generates pulsed discrete/sampled signal. So, unlike in natural sampling, the sampled signal remains constant for a while in flat top sampling. In natural sampling, the sampled signal follows the actual signal values for the duration when the sampled signal is non-zero. An example of flat top sampling is shown in Figure 4.1(c).

4.4.3 Frequency Domain Analysis of Sampling

Sampling can be mathematically represented as a process of multiplying two signals. In ideal sampling, the signal to be sampled is multiplied with impulse train to obtain sampled signal. We consider this case for frequency domain analysis of sampling. Through further discussions on sampling, we will establish the following:

- Minimum sampling rate
- Relation between minimum sampling rate and signal bandwidth
- Sampling theorem and Nyquist rate
- Concept of aliasing

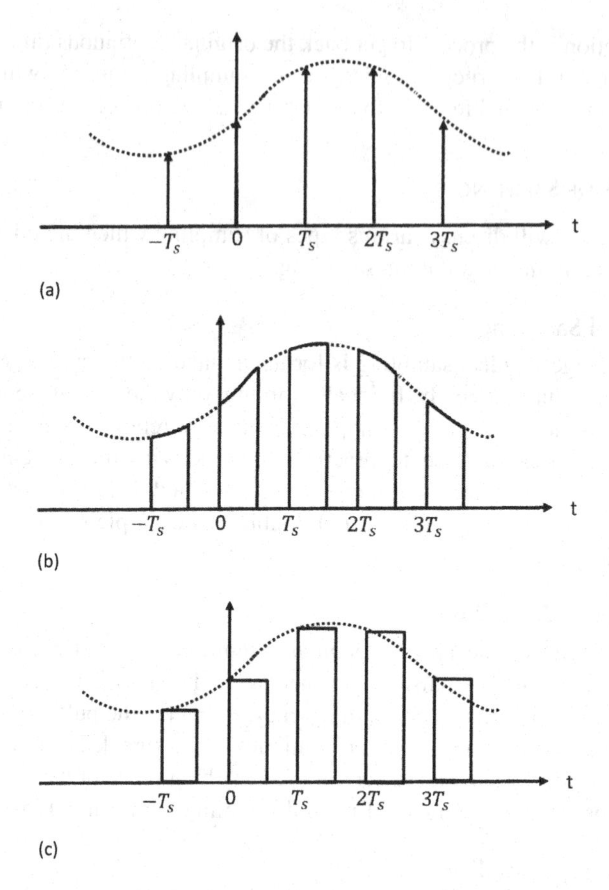

FIGURE 4.1 Signal sampling using various types for the original signal shown with a dotted line: (a) Ideal sampling: Samples are shown with impulses at sampling instants. (b) Natural sampling: The pulses of predefined width start at sampling instant and follow the envelope of the signal. (c) Flat top sampling: The pulses of predefined width start at sampling instant with the same amplitude of signal at the sampling instant.

Consider a bandlimited signal, $g\ (t)$, with bandwidth, B Hz, to be sampled. The frequency domain representation of the signal is shown in Figure 4.2(a). The signal is multiplied with an impulse train represented as

$$\delta_{T_s}(t) = \sum_{n=-\infty}^{\infty} \delta(t - nT_s) \tag{4.3}$$

where $\delta(\cdot)$ is Dirac delta function. The impulse train represented in the above expression is a periodic signal; therefore, it can be represented by Fourier series. The Fourier series of the signal represented by (4.3) can be represented as

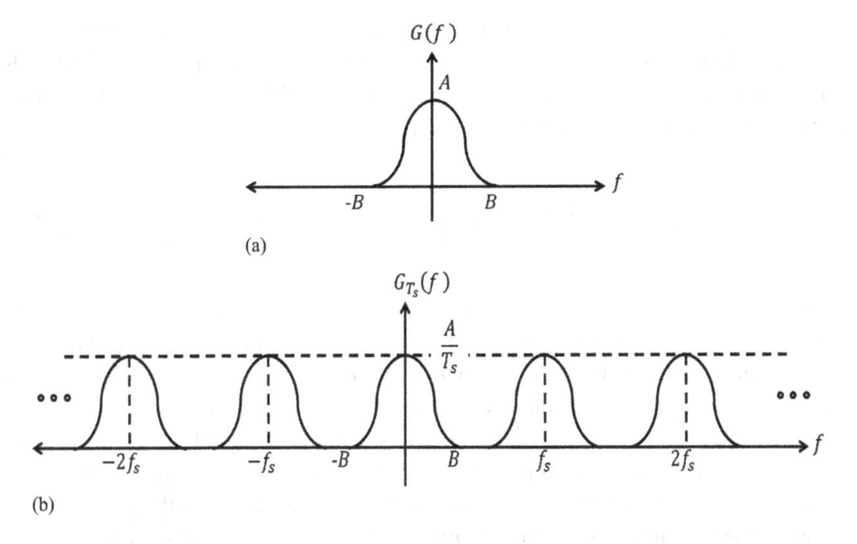

FIGURE 4.2 Frequency spectrum of signal before and after sampling: (a) Frequency domain representation of the signal to be sampled, (b) Frequency spectrum of sampled signal with sampling frequency, f_s.

$$\delta_{T_s}(t) = \frac{1}{T_s}\sum_{n=-\infty}^{\infty} e^{jn\omega_s t} \tag{4.4}$$

where $\omega_s = 2\pi f_s$ is given by $\omega_s = \frac{2\pi}{T_s}$ and $f_s = \frac{1}{T_s}$ is the sampling frequency in Hertz or samples per seconds.

The sampled signal can be represented as

$$\begin{aligned} g_{T_s}(t) &= g(t)\delta_{T_s}(t)s \\ &= \sum_{n=-\infty}^{\infty} g(nTs)\delta(t - nTs) \end{aligned} \tag{4.5}$$

Using (4.4) in the above expression, the sampled signal can be alternatively represented as

$$g_{T_s}(t) = \frac{1}{T_s}\sum_{n=-\infty}^{\infty} g(t)e^{jn\omega s^t} \tag{4.6}$$

Note that the above signal representation can be compared with the DSB-SC signals discussed in the previous chapter (see expression (3.3)), i.e., the sampled signal can be represented as a sum of the message signal and DSB-SC modulated signals with carrier frequency, ω_s, and its harmonics. So, the frequency domain representation of sampled signal represented by (4.6) follows from the concepts we already discussed for frequency domain representation of DSB-SC signals. Thus, the Fourier transform of the sampled signal, $G_{Ts}(f)$, can be given as

$$G_{T_s}(f) = \frac{1}{T_s}\sum_{n=-\infty}^{\infty} G(f - nf_s) \tag{4.7}$$

where $G(f)$ is the frequency spectrum (Fourier transform) of the signal, $g(t)$. The expression (4.7) represents the sum of frequency spectrum of message signal, $g(t)$, and the spectra shifted in frequency by harmonics of sampling frequency. The spectra of the signal, $g(t)$, and the sampled signal are shown in Figure 4.2.

Further, the Fourier domain representation of sampled signal also follows from the properties of Fourier transform and Fourier series, i.e., if two signals are multiplied in time domain, the Fourier transform/series representation of the product is the convolution between the Fourier transform/series of the individual signals.

After sampling, the following changes in signal characteristics are important to note.

- A bandlimited (finite bandwidth) signal becomes an infinite bandwidth signal after sampling.
- One part of the spectrum of sampled signal is the spectrum of the original signal itself, i.e., the term corresponding to $n = 0$ in expression (4.7).
- By selecting a proper filtering mechanism, the original signal can be recovered if proper T_s is selected while sampling, for example, lowpass filter for a baseband signal, $g(t)$.

4.4.4 Nyquist Rate and Sampling Theorem

Based on the frequency spectrum that we obtain after the sampling, we will now find the minimum sampling frequency that would enable us to reconstruct the original signal from the samples. For this, the frequency spectrum of sampled signal with different sampling frequencies is shown in Figure 4.3. We discuss these cases one by one.

Case I ($f_s > 2B$): In this case, it is observed that the spectrum of different spectral components like original signal and modulated signals at the harmonics of sampling frequency do not overlap as seen in Figure 4.3(a). The different components are separated well and the original signal can be recovered through lowpass filtering with cutoff frequency of B Hz.

Case II ($f_s = 2B$): In this case, it is observed that the spectrum of different spectral components like original signal and modulated signals at the harmonics of sampling frequency do not overlap as seen in Figure 4.3(b). The different components are still separated such that the original signal can be recovered through lowpass filtering with sharp transition at frequency of B Hz from passband to stop band. This needs an ideal lowpass filter.

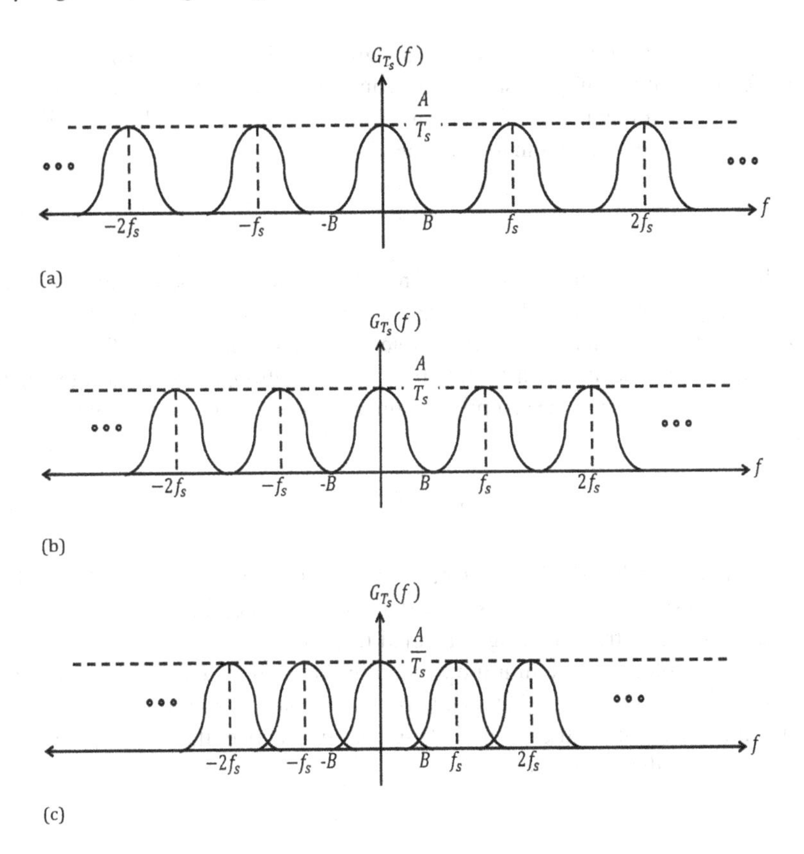

FIGURE 4.3 Frequency spectrum of sampled signal with different sampling frequencies: (a) Sampling frequency more than twice the bandwidth of signal to be sampled, $f_s > 2B$. (b) Sampling frequency equal to twice the bandwidth of signal to be sampled, $f_s = 2B$. (c) Sampling frequency less than twice the bandwidth of signal to be sampled, $f_s < 2B$.

Case III ($f_s < 2B$): In this case, it is observed that the spectrum of different spectral components like original signal and modulated signals at the harmonics of sampling frequency overlap as seen in Figure 4.3(c). The different components like original signal or the modulated signals at harmonics of sampling frequency cannot be separated using filtering. So, the original signal cannot be reconstructed when $f_s < 2B$. Such a sampling is referred to as 'under-sampling'. If the sampled signal is passed through a lowpass filter with cutoff frequency of B Hz, the original signal and the folded tail of overlapping part are reconstructed, i.e., original signal cannot be recovered in this case.

Based on the above discussions, it is concluded that a bandlimited signal with bandwidth of B Hz must be sampled with sampling frequency which is at least twice the bandwidth of the signal to be sampled. This minimum sampling frequency of $2B$ Hz is known as 'Nyquist rate'. On a similar note, Nyquist interval is defined as $\frac{1}{2B}$

which is the largest sampling interval for a signal with bandwidth B Hz to be recovered perfectly from its sampled signal. The same conclusion of the above discussion is known as a 'sampling theorem'. The sampling theorem for a bandlimited signal of bandwidth B Hz can be mathematically stated as

$$f_s \geq 2B \tag{4.8}$$

Further, the above condition is valid only if $G(B) = 0$ in a baseband signal. For example, let $g(t) = \cos(2\pi Bt)$. The frequency domain representation of $g(t)$ contains an impulse at $f = B$. In this case, if we sample the signal at the rate of $2B$ samples per second, there would be an overlap in the frequency spectrum of sampled signal at the locations $\pm B, \pm 2B, \cdots$. So, the sampling frequency must satisfy the following in such cases.

$$f_s > 2B, \tag{4.9}$$

i.e., the equality condition of expression (4.8) no longer exists for signals in which $G(B) /= 0$. Such signals have to be sampled always at a rate higher than the Nyquist rate in order to reconstruct the original signal from the samples.

It should be noted that the minimum sampling rates that have been mentioned in this section are the ideal rates that enable perfect reconstruction of the original continuous time signals from their sampled signals. In practice, the sampling rates are much higher than the Nyquist rates.

Here, we have the following remarks to make about sampling of baseband signals and passband signals:

- Highest frequency component and bandwidth are the same for a baseband signal.
- Highest frequency component is different from the bandwidth for a passband signal.
- Nyquist rate for the baseband signal can be defined in terms of highest frequency component for a baseband signal.
- Nyquist rate for the passband signal can only be defined in terms of bandwidth for a passband signal.

4.4.5 Aliasing

We already discussed that if $f_s < 2B$, the repetitions of $G(f)$ at its own location and multiples of f_s will overlap as shown in Figure 4.3(c). This overlap is known as 'aliasing'. The process of aliasing is of practical importance. This is because all the practical signal are band unlimited, i.e., they exist for an unlimited bandwidth. However, the significant part of these signals can be represented in finite bandwidth. This is required to have a finite sampling frequency and a perfect reconstruction from the sampled signals.

Why are the practical signals of infinite bandwidth?

- All practical signals exist for a finite time; for example, a phone call exists only for a while.
- Such signals are considered as multiplication of the signal itself and a rectangular pulse that exists for the duration of the existence of the signal.
- The spectrum of rectangular pulse of finite duration is a sinc function which has infinite bandwidth although maximum energy is contained in a finite bandwidth.
- The resulting signal is of infinite bandwidth.
- All finite duration signals are of infinite bandwidth.
- Signals with finite duration and finite bandwidth do not exist in practice.

To avoid aliasing due to sampling of practical signals, the signals are passed through antialiasing filter before sampling. Antialiasing filter is a lowpass filter whose cutoff frequency is set as $\frac{f_s}{2}$. The cutoff frequency of antialiasing filter and sampling rate are decided such that no significant information of the signal is lost.

4.4.6 Reconstruction of Signal from Samples

We have already established the fact that if sampling frequency is selected above the Nyquist rate, the original signal can be recovered without loss of information from the sampled signal. It is also mentioned that the sampled signal has to pass through a lowpass filter with signal bandwidth as cutoff frequency in case of baseband signals.

Consider the case of ideal sampling and ideal lowpass filtering. The frequency response of ideal lowpass filter is given by a rectangular pulse centered at zero frequency and width as double the cut-off frequency and is shown in Figure 4.4. The impulse response, $h(t)$, of the ideal lowpass filter can be obtained by using the properties of Fourier transform and is given as

$$h(t) = \text{sinc}(2\pi Bt). \tag{4.10}$$

Using the properties of linear time invariant systems, the output of filter can be given as

$$g(t) = \sum_k g(kTs)h(t - kTs) \tag{4.11}$$

where $g(kT_s)$ is the kth sample of original continuous time signal which is the instantaneous value of signal at $t = kT_s$. Note that the kth sample of original signal is an impulse at the location $t = kT_s$ with magnitude $g(kT_s)$. So, the signal recovered only

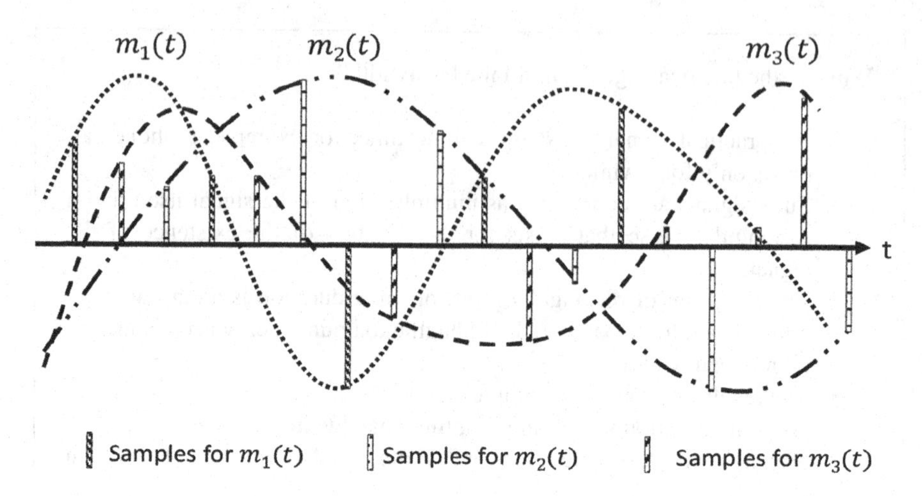

FIGURE 4.4 Demonstration of time division multiplexing of samples from three signals.

from the kth sample is $g(kT_s)\,h(t - kT_s)$. Since the sampled signal is the sum of all the impulses at location $t = kT_s$ with different values of k, the reconstructed signal is a summation of responses to all such impulses as represented in (4.11). The reconstructed signal can further be represented as

$$g(t) = \sum_k g(kTs)\text{sinc}(2\pi B(t - kTs)) \tag{4.12}$$

The above expression is known as an 'interpolation formula'. The name comes from the fact that the entire signal is obtained from a set of discrete values at uniform intervals. Further, $Ts = \frac{1}{2B}$ when the sampling rate is the same as the Nyquist rate, the interpolation formula is further simplified as

$$g(t) = \sum_k g(kTs)\text{sinc}(2\pi Bt - k\pi) \tag{4.13}$$

4.4.7 Time Division Multiplexing

Time division multiplexing is the technique of transmitting signals from multiple sources over a single line without interference among them. Using sampled signals, we can easily transmit various signals through a single line using time division multiplexing. Consider three signals are to be transmitted over a single line. The signals are sampled at a rate of F_s samples per second. So, we collectively get $3F_s$ samples per second for all three signals. All three of these signals can be transmitted over a single line having capacity to transmit $3F_s$ samples per second. This can be done by selecting a sample from one signal at a time,i.e., serially. The sample selection methodology for time division multiplexing of three signals is demonstrated in Figure 4.4.

4.4.8 Quantization

Quantization is the process that actually converts the signal from analog to digital. After sampling, we get a discrete time signal, but it is still an analog signal. In quantization, the magnitude axis is divided into a certain specified number of levels. Each level is assigned a certain range. All the signal values within this range are assigned the same level. Thus, quantization is referred to as a process of converting a discrete magnitude axis. Quantization can be classified into two types: (1) uniform quantization and (2) non-uniform quantization.

Here, we define some terms related to the process of uniform quantization.

- Quantization range: It is the range of the message signal values in which the quantizer works. Let us assume the message signal is $g(t)$. The range $(-g_p, g_p)$ means that any values of message below $-g_p$ and above g_p will be chopped off to $-g_p$ and g_p, respectively. So, the range is a parameter of quantizer and does not specify a limit on the message signal values.
- Number of levels (L): The entire quantization range is divided into uniform intervals. The number of such intervals is referred to as the number of levels, L.
- Step size, Δv: It is the size of an interval when the quantization range $(-g_p, g_p)$ is divided into L levels of equal size. The step size can be given as

$$\Delta v = \frac{2g_p}{L} \tag{4.14}$$

  Further, the quantized level is considered the midpoint of the interval in which the sample value lies.
- Quantization noise: Quantization is an irreversible process. When the original continuous time signal is reconstructed from quantized samples, an error due to quantization is always present. This error is known as 'quantization noise'. Quantization noise has direct relation with the number of levels, L. As L increases, quantization noise reduces because of reduction in step size.

We now analyze the quantization noise power and the degradation in signal quality due to the quantization error. The reconstruction of message signal from sampled signal and quantized signal can be represented in the form of interpolation formula as

$$g(t) = \sum_k g(kTs)\,\mathrm{sinc}(2\pi Bt - k\pi) \tag{4.15}$$

and

$$\hat{g}(t) = \sum_k \hat{g}(kTs)\,\mathrm{sinc}(2\pi Bt - k\pi) \tag{4.16}$$

where $g(t)$ and $\hat{g}(t)$ are the reconstructed signals from samples before quantization and after quantization, respectively, and $g(kT_s)$ and $\hat{g}(kT_s)$ are k^{th} samples of signal before quantization and after quantization, respectively.

The quantization noise can be written in terms of reconstructed signal and respective samples as well. The quantization noise signal can be given as

$$\begin{aligned} q(t) &= \hat{g}(t) - g(t) \\ &= \sum_k \left[\hat{g}(kTs) - g(kTs)\right] \operatorname{sinc}(2\pi Bt - k\pi) \end{aligned} \tag{4.17}$$

while the quantization noise in k^{th} sample can be given as

$$q(kTs) = \hat{g}(kTs) - g(kTs) \tag{4.18}$$

Now, we use the theory of probability to calculate the power contained in quantization noise signal or quantization noise samples. We recall that the step size is $\Delta v = \frac{2g_p}{L}$ and the quantized samples are valued as the midpoint of the interval in which the original sample lies. So, the range of quantization error is $(-\Delta v/2, \Delta v/2)$. Further, we assume that the quantization error is uniformly distributed in the range $(-\Delta v/2, \Delta v/2)$. So, the probability density function (PDF) of quantization noise can be given as

$$p_x(q) = \frac{1}{\Delta v} \tag{4.19}$$

for $-\Delta v/2 < 0 < \Delta v/2$, and zero otherwise. With such statistical properties, the quantization noise has zero mean value, i.e., $\bar{q} = 0$. Thus, quantization noise power, N_q, can be given as the mean square value of the quantization noise signal which is the same as its variance which can be calculated as

$$\begin{aligned} N_q = \bar{q}^2 &= \int_{-\Delta v/2}^{\Delta v/2} \frac{q2}{\Delta v}\, dq \\ &= \frac{(\Delta v)^2}{12} \\ &= \frac{g^2{}_p}{3L^2} \end{aligned} \tag{4.20}$$

It is clear from the above expression that quantization noise power has a direct dependence on step size. Further, step size depends on the quantization range and number of levels. So, quantization noise power increases with increase in step size and reduces with increased number of levels for a given quantization range.

After quantization, we get a digital signal (although not in binary form). So, quantization is the actual process where analog signal is converted into a digital signal. Further, we discuss the binary representation of the digital signal obtained after quantization.

4.4.9 Encoding and Pulse Code Modulation

The quantization process assigns one of the fixed values that belongs to L levels of quantization. These fixed values are mid-point of each interval. So, a signal will have either of L values at any time after the quantization. Now, these fixed values have to be encoded into binary codes of length $n = \log_2 L$. Table 4.1 shows the values of range of samples corresponding to different levels, the quantized value corresponding to that level and binary code mapping. Note that the values assumed for preparing this table are as follows.

- Quantization range: $(-g_p, g_p) = (-1,1)$, i.e., $g_p = 1$. Any sample values above 1 and below –1 are clipped to 1 and –1, respectively.
- Number of levels, $L = 8$. These levels are counted from 0 to 7.
- Step size, $\Delta v = 0.25$. The step size is obtained as $\Delta v = \dfrac{2g_p}{L}$.

The encoding is a process of converting the quantized sample values in their binary codes. This process is known as pulse code modulation (PCM). Recall the definition of 'modulation' as the process in which some parameter of carrier signal varies as per the instantaneous magnitude of the message signal. In PCM, the sequence of quantized samples is considered as a message signal, the sequence of binary digits (bits) is considered as a carrier signal, and the sequence of bits is changed as per the instantaneous value of the quantized sample sequence. So, PCM justifies being a modulation scheme. However, it does not translate the frequency spectrum of the message signal from baseband to passband. After PCM, the signal is still a baseband signal.

Furthermore, after encoding the quantized sample values into a binary code, the bits may be transmitted serially (one by one over a single line) or in parallel (all at a time over multiple lines) fashion for further processing. Also, there are multiple ways to represent bit '0' and bit '1' while transmission. The scheme for representation of

TABLE 4.1
An Example of Encoding: Mapping of Quantized Sample Values to Binary Code for $L = 8$, $n = 3$, and Range (–1,1)

Level	Range of Sample Values	Quantized Sample Value	Binary Code
0	–1 to –0.75	–0.875	000
1	–0.75 to –0.5	–0.625	001
2	–0.5 to –0.25	–0.375	010
3	–0.25 to 0	–0.125	011
4	0 to 0.25	0.125	100
5	0.25 to 0.5	0.375	101
6	0.5 to 0.75	0.625	110
7	0.75 to 1	0.875	111

bits while transmission over line is commonly known as a 'line code'. We will discuss some line encoding schemes in the later sections of this chapter.

4.5 DIFFERENTIAL PCM

It is already established that the quantization noise power and the step size are related by a square low. The exact relation between quantization noise poser and the step size is

$$N_q = \frac{(\Delta v)^2}{12}. \tag{4.21}$$

4.5.1 Reduction of Quantization Noise

There are two ways of reducing the step size and hence the quantization noise: (1) reduction in the range of quantizer, i.e., g_p, and (2) increase in the number of levels, L.

Each method will have certain limitations on signal handling capacity of the system.

- Reduction in the range of quantizer will lead to more distortions when we convert the signal back to its original form. The reason behind this is the clipping off of the signal beyond the range of values, i.e., below $-g_p$ to $-g_p$ and above g_p to g_p. Reducing g_p means more and more samples will be mapped to the highest and lowest levels. This will bring larger error in the reconstructed signal or in the signal when converted back to analog signal.
- Increasing the number of levels has a direct impact on the code length because the number of bits required would be larger to encode a larger number of levels. Further, a large number of bits per sample means larger bandwidth requirement for transmission of these bits (when transmitted serially) or more parallel lines (when transmitted in parallel).
- For a scheme in which each sample is encoded with a codeword of n bits, the rate at which information transmission to be done is $2nB$ bits per second when the signal of bandwidth B Hz is sampled at Nyquist rate. So, $2nB$ bits per second is the minimum bit rate. This will be higher for practical scenarios where the required sampling frequency is higher than the Nyquist rate.

4.5.2 Concept of Differential Sampling

The required sampling rates are much higher than the Nyquist rate for practical systems. So, the successive samples are expected to have similar values. Considering this fact that practical sampling rates are higher than the Nyquist rate and the successive samples are of similar values, a technique of differential sampling is thought of as a method to reduce the signal amplitude range. Once we reduce the signal amplitude range, we can safely reduce the quantization range by reducing g_p such that $g_p > \max(g(t))$ where $g(t)$ is the signal to be sampled.

Now, we denote the kth sample of the signal, $g(t)$ at time $t = kT_s$ as $g[k]$. This is to differentiate the continuous time signals and discrete time signals. As already discussed, the transmission of samples is in the form of difference of consecutive samples. So, the kth differential sample, $d[k]$ at time $t = kT_s$ to be transmitted can be given by

$$d[k] = g[k] - g[k-1]. \tag{4.22}$$

Further, having the information about $d[k]$ and $g[0]$, the original samples can be obtained iteratively as

$$\begin{aligned} g[1] &= d[1] + g[0] \\ g[2] &= d[2] + g[1], \\ &\vdots \end{aligned} \tag{4.23}$$

i.e., the first sample, $g[0]$, is required in its original form. The idea of differential sampling is that the peak value of $d[k]$ is generally much smaller than that of the original samples. So, we can reduce the range of the quantizer to reduce step size or bandwidth can be saved by reducing both the number of levels and the quantization range.

The concept of differential sampling can be summarized as follows:

- The sampling frequency is set much higher than the Nyquist rate to reduce the difference between successive samples.
- The difference between successive samples is taken. The signal received after taking the difference is referred to as 'differentially sampled signal'.
- The differentially sampled signal usually has lesser amplitude range than that of the original signal.
- There can be two ways of taking advantage from the reduced amplitude range:
  1. The quantization range can be reduced to reduce the quantization noise power.
  2. The number of levels can be reduced to reduce the transmission bandwidth of the signal.
- The differential samples received after taking the difference of successive samples are quantized.
- The quantized samples are then encoded with corresponding binary codes of appropriate length.
- This complete process is known as 'differential PCM' (DPCM) because the differences of successive samples are quantized and encoded for conversion of analog signals to digital signals.

We discussed the successive samples' difference based on DPCM in this section. The performance of DPCM can be further improved using predicted values of kth

sample using the values of past samples. We will establish the concepts behind prediction-based DPCM in the next section.

4.5.3 Prediction-Based DPCM

We first discuss the concept of prediction in brief before going into the discussion of prediction-based DPCM.

4.5.3.1 Concept of Prediction

The theory of prediction is based on the fact that the future values of a signal or sequence can be represented as a function of its past and present values. One of the examples is Taylor series representation of functions. Using Taylor series expansion, a function can be represented as

$$f(x+\Delta x) = f(x) + \frac{\Delta x}{1!} f'(x+\Delta x) + \frac{\Delta x^2}{2!} f''(x+\Delta x) + \cdots \tag{4.24}$$

where the representation can be truncated up to first derivative term if the value of Δx is sufficiently small. Further, the above expression can be extended to a discrete signal/function case by replacing the differentiation terms by difference terms of the same order.

So, if we consider T_s to be sufficiently small such that $(k + 1)^{th}$ sample can be predicted using Taylor series truncated up to first order difference, then $(k + 1)^{th}$ sample can be given as

$$\hat{g}[k+1] = g[k] + TsD\{g[k]\} \tag{4.25}$$

where $\hat{g}$ is the predicted value of g and $D\{g[k]\}$ is the first order difference of $g[k]$ given as

$$D\{g[k]\} = \frac{g[k] - g[k-1]}{Ts}, \tag{4.26}$$

i.e., the $(k + 1)^{th}$ sample value as predicted by expression (4.25) uses k^{th} sample and $(k + 1)^{th}$ sample. Further, the accuracy of prediction can be improved by using higher orders of differences in the prediction expression. The larger the order of differences used for prediction, the larger is the number of past samples required, and the better is the accuracy of prediction. In general, the prediction expression using N samples can be given as

$$\hat{g}[k] = a_1 g[k-1] + a_2 g[k-2] + a_3 g[k-3] + \cdots + a_N g[k-N], \tag{4.27}$$

i.e., $\hat{g}[k]$ can be represented as a linear combination of past N samples. This kind of predictors is known as 'linear predictors' or 'linear prediction filters', where

$a_1, a_2, \cdots, a_N$ are known as the 'coefficients of prediction filters'. The evaluation of coefficients of prediction filters is out of the scope of this book. Interested readers may refer to [1] for detailed discussions on prediction filters as well as other types of adaptive filters.

4.5.3.2 DPCM Based on Prediction Theory

The concept of DPCM is based on transmission of differences in place of original samples to reduce the magnitude range so that the range of quantizer can be reduced for advantages like reduction in quantization noise power and/or reduction in transmission bandwidth. Using the prediction theory, we already show that the samples can be predicted using past samples. In prediction-based DPCM, we use the difference of sample and its predicted value for transmission which can be given as

$$d[k] = g[k] - \hat{g}_q[k] \tag{4.28}$$

where $\hat{g}_q[k]$ is the kth sample value predicted from the quantized samples, i.e. the prediction of present sample is done using quantized past sample values. Further, the quantized differential samples can be given as

$$d_q[k] = d[k] + q[k] \tag{4.29}$$

where $q[k]$ is the quantization noise added in the quantization process.

Now, we establish that the quantization noise, $q[k]$ that reflects in the differential samples, $d[k]$ is the same as that would be added to the original samples. Further, we know that the quantized sample values can be represented as the sum of original sample and the quantization noise shown below,

$$g_q[k] = g[k] + q[k]. \tag{4.30}$$

Using the expressions (4.28), (4.29), and (4.30), we can show that the quantized signal samples can be given as

$$g_q[k] = \hat{g}[k] + d_q[k] \tag{4.31}$$

which shows that the input to the predictor should be the sum of predicted sample value and the quantized differential samples as shown in Figure 4.5.

Now, let's discuss recovering the signal samples from the differential samples, $d_q[k]$, which are transmitted. For recovering the original samples, we assume that the predictors used at the transmitter and the receiver are the same, i.e., the input to the predictor is quantized signal samples. So, the signal recovery at the receiver can be done by the following steps:

- Input or the received sequence is quantized differential samples, $d_q[k]$.
- Predictor at the receiver predicts the quantized signal samples, $\hat{g}_q[k]s$.

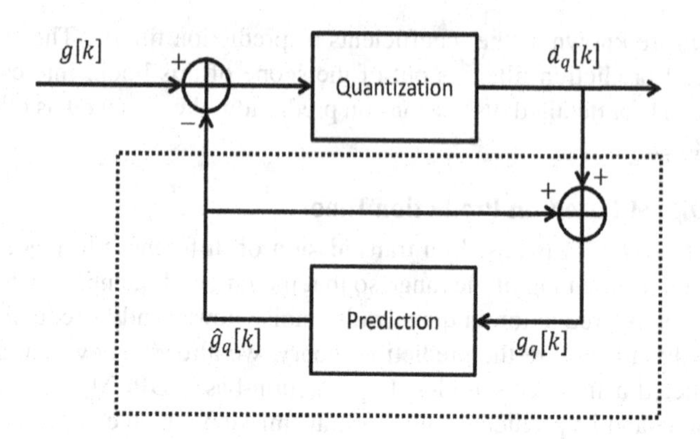

FIGURE 4.5 Block diagram of differential PCM system. The part shown in the dotted rectangle is the receiver system for DPCM.

- Add predicted quantized signal samples to the received samples to get the quantized signal samples,

$$g_q[k] = d_q[k] + \hat{g}_q[k]. \tag{4.32}$$

- Use the above addition as input to the predictor for prediction of the next quantized sample value.

In addition to our discussions of analog to digital signal conversions, we will discuss one more method which is the delta modulation. Delta modulation is a further modification of DPCM.

4.6 DELTA MODULATION

As mentioned in earlier sections, differential sampling is of great importance to improve the performance of analog to digital signal conversion by reducing quantization noise or transmission bandwidth. In this section, we will discuss one more technique which uses differential sampling for analog to digital signal conversion: delta modulation. Delta modulation is a special case of DPCM in which each differential sample is encoded into a single bit, i.e., the differential sample is transmitted in the form of bit '0' or bit '1' in delta modulation. In simple words, delta modulation can be referred to as '1-bit DPCM' or 'single bit DPCM'. In addition, the prediction design is also simple as compared to DPCM in delta modulation. The differential sample in delta modulation is represented as

$$d_q[k] = g_p[k] - g_q[k-1], \tag{4.33}$$

i.e., difference of successive quantized samples which in original DPCM without predictor was the difference between successive samples before quantization. So, delta modulation can be considered as prediction-based DPCM in which the predicted quantized sample value is the previous quantized sample value itself, i.e., first order prediction or the easiest prediction method.

Further, the kth quantized sample can be obtained iteratively using the differential quantized samples using the expression (4.33) as

$$g_q[k] = \sum_{n=0}^{k} d_q[n]. \tag{4.34}$$

Thus, we can summarize the properties of delta modulation as follows:

- Predictor in the DPCM system can be replaced by a delay equivalent to the sampling interval to obtain a delta modulation system.
- Quantizer in a delta modulation system is a comparator.
- The receiver for a delta modulation system can be designed as a simple accumulator that adds the incoming signal into the sum of all samples received earlier.

4.7 LINE ENCODING SCHEMES

After the conversion of an analog signal into a digital signal, the signal is represented in the form of bits. This bit represented digital signal has to be processed further and transmitted over wired or wireless medium to the end user, say, receiver. For the purpose of transmission, this bit sequence has to be converted into an electrical signal in the form of some voltage waveform. Usually, the bit levels are represented as certain voltage levels like 0 volts to bit '0' and 5 volts to bit '1' in TTL (transistor transistor logic) families. In a similar way, there can be different ways to represent waveforms for these bits over transmission lines. The scheme for representation of bits while transmission over line is commonly known as 'line code' and the process of such encoding is known as 'line encoding'.

There are mainly three types of line encoding schemes: (1) unipolar, (2) polar, and (3) bipolar.

Further, in each of the above three schemes there are two subcategories as follows:

1. Non-return to zero (NRZ): In NRZ encoding schemes, the pulse does not return to zero within the bit duration.
2. Return to zero (RZ): In RZ encoding schemes, the pulse exists only for the half-bit duration. Other half-bit duration is represented with a pulse of magnitude zero.

4.7.1 UNIPOLAR ENCODING

In unipolar scheme of line coding, the pulses to represent bit '1' are with positive polarity, and the representation for bit '0' is without pulse. For this reason, a polar

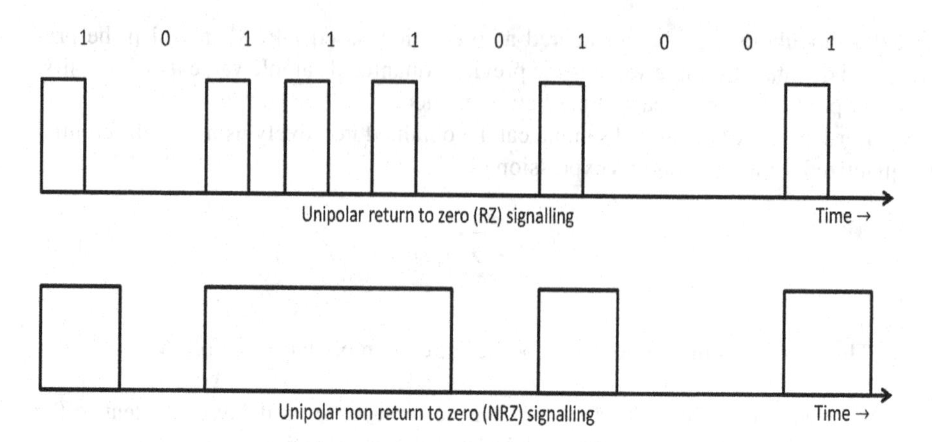

FIGURE 4.6 An example of unipolar signalling with return to zero (RZ) and non-return to zero (NRZ) type of encoding.

signalling scheme is also known as 'on-off keying scheme'. So, only single polarity pulses occur in unipolar signalling schemes. An example of unipolar encoding scheme with RZ and NRZ types of encoding is shown in Figure 4.6 for a 10-bit sequence.

4.7.2 Polar Encoding

In polar scheme of line coding, the pulses to represent bit '1' are with positive polarity and to represent bit '0' is with negative polarity pulse of the same magnitude as that of positive polarity pulse that represents bit '1'. So, both positive and negative polarity pulses occur in polar signalling schemes. The average value of signal with polar encoding is zero when bit '0' and bit '1' occur with equal probability of 0.5. So, DC component does not exist in the spectrum of polar signalling. An example of polar encoding scheme with RZ and NRZ types of encoding is shown in Figure 4.7 for a 10-bit sequence.

4.7.3 Bipolar Encoding

Like unipolar scheme, the representation for bit '0' is without pulse in bipolar encoding scheme. But bit'1' is represented with both positive and negative polarity pulses of the same magnitude. The polarity of pulse is reversed on each occurrence of bit '1'. Like polar signalling, the average value of signal with bipolar encoding is zero. So, DC component does not exist in the spectrum of bipolar signalling. An example of bipolar encoding scheme with RZ and NRZ types of encoding is shown in Figure 4.8 for a 10-bit sequence.

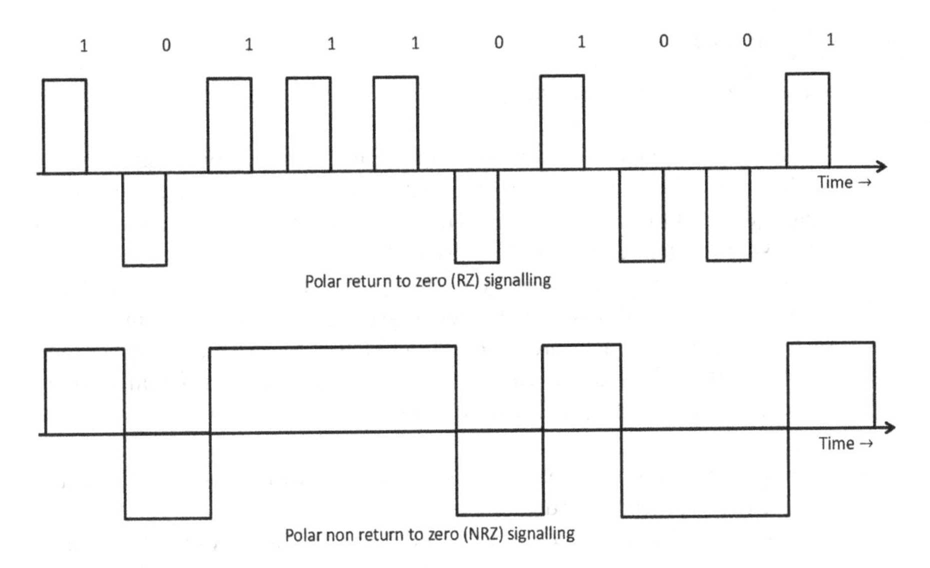

FIGURE 4.7 An example of polar signalling with return to zero (RZ) and non-return to zero (NRZ) type of encoding.

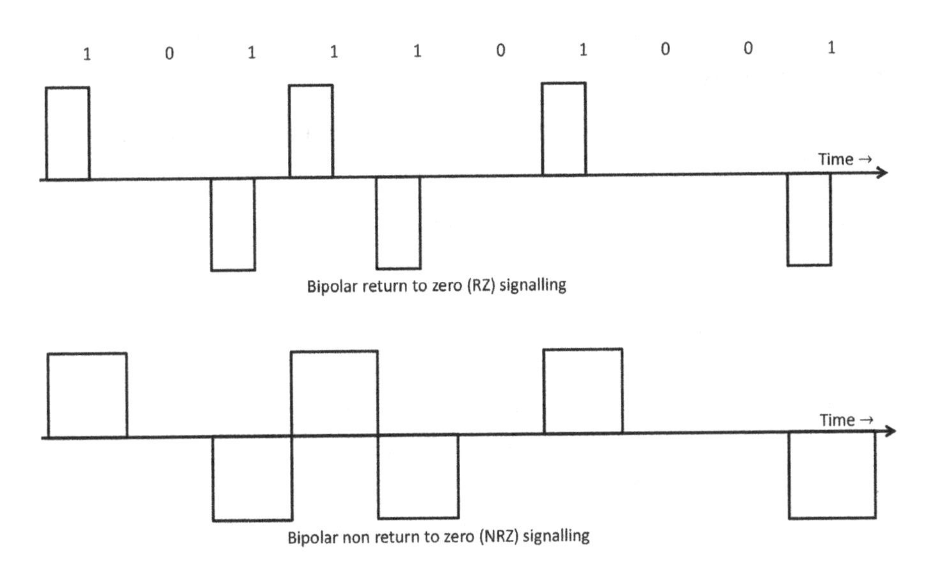

FIGURE 4.8 An example of bipolar signalling with return to zero (RZ) and non-return to zero (NRZ) type of encoding.

4.8 SUMMARY

In this chapter, we note the following points:

- Digital systems reduce hardware complexity and improve noise immunity.
- Analog signals can be converted to digital signals through processes like sampling, quantization, and encoding. This whole sequence of analog to digital conversion is often referred to as 'pulse code modulation'.
- The minimum sampling frequency must be twice the signal bandwidth (also called 'Nyquist rate') so as to enable perfect reconstruction of the original signal.
- Since the practical signals are band unlimited, they are passed through a low-pass filters (known as 'antialiasing filters') such that no significant information is lost and the signal becomes bandlimited. This is to avoid aliasing in the sampled signal.
- Quantization is an irreversible process. It adds quantization noise to the signal which can be reduced by reducing the step size.
- The higher the number of levels in quantization, the larger would be the data rate after PCM, and hence the larger would be the transmission bandwidth required.
- Differential PCM is an effective way to reduce the quantization noise and the transmission bandwidth when sampling rate is sufficiently high.
- When sampling rates are very high, delta modulation is a highly effective technique to convert an analog signal to digital (binary) signal.

BIBLIOGRAPHY

1. Haykin, Simon. *Adaptive Filter Theory*, 4th ed., Prentice Hall, Upper Saddle River, NJ, 2002.

5 Digital Communication Systems

5.1 INTRODUCTION

As in the case of analog communication systems, electromagnetic waves carry the information from transmitter to receiver in digital communication systems as well. As discussed in the previous chapter, we will consider the digital form of a message signal in digital communication systems. Let us introduce a new class of systems here. There are systems in which the message signal is in the analog form, but the carrier signal is in the digital form. Such systems are referred to as 'pulse based' systems because of the pulsed form of the carrier signal.

There are various classifications of communication systems based on various parameters as follows:

1. Classification based on the medium of transmission
   a. Wired communication: Communication in which information is carried by wires. The wires may be of various types like twisted pair, parallel conductors, center core cable, optical fiber, etc.
   b. Wireless communication: Communication in which information is transmitted to the receiver over free space in wireless communication. This type of communication systems includes cellular communication, mobile communication, satellite communication, etc.
2. Classification based on the type of message signal
   a. Analog communication: Transmission of message signals in analog form
   b. Digital communication: Transmission of message signals in digital form
3. Classification based on the frequency of operation
   a. Baseband communication: Transmission of signals without modulation, i.e., spectrum centered around zero frequency
   b. Bandpass communication: Transmission of modulated signals with spectrum translated to carrier frequency

We have already discussed the differences between analog communication systems and digital communication systems. In this chapter we start our discussion with a brief description of baseband and passband communication systems. Further, we will discuss pulse based communication systems in which a carrier is in digital form (pulsed form) and a message is a continuous time analog signal. The discussion on pulse based systems is followed by carrier modulation in digital communication systems. We will discuss the effect of noise on a constellation diagram of digital modulation schemes. We will also introduce a new class of modulation scheme multiplexing named 'spread spectrum technique' which is the technology behind code division multiple access (CDMA) standard for wireless communication.

DOI: 10.1201/9781003213468-5

5.2 BASEBAND AND PASSBAND COMMUNICATION SYSTEMS

Baseband and passband communication systems are classified based on the spectrum of the signal to be transmitted. We know the steps of processing required for converting any continuous time analog signal into discrete time analog signal through sampling and further converting the discrete time analog signal to digital signal through quantization. Further, we get a binary form of a digital signal through pulse code modulation (PCM), and various types of line encoding schemes represent the signal in pulsed form which finally gets a continuous time digital signal. These signals, whether they are original continuous time analog signals or digital signals obtained after line encoding, would have similar frequency spectra in one aspect. That one aspect is the spectrum concentrated near zero frequency. So, the signal is a baseband signal irrespective of being in analog form or its corresponding digital/binary form represented by a suitable line encoding scheme.

5.2.1 Baseband Communication Systems

Baseband communication systems are usually in the form of pulsed transmission of messages. Mainly, when digital messages are transmitted in binary form, the bits are encoded and represented in the form of electrical signal using one of the line encoding schemes discussed in the previous chapter. These line-encoded bit sequence signals are transmitted over a wired connection between transmitter and receiver.

As already discussed, the spectrum of message signal is concentrated around zero frequency. So, multiple messages sent over a single connection would cause interference among the messages. Therefore, a dedicated line connection is required between transmitter and receiver for this purpose to avoid interference. One way of transmitting multiple message signals over a single line is time division multiplexing in baseband communication systems. In baseband communications, the signals are transmitted without frequency translation, i.e., carrier based modulation schemes are not required in baseband communication systems.

It should be noted that even analog signals can be used for transmission in baseband communication systems. In case the message signal is an analog signal and needs to be transmitted without conversion into a digital signal, the line encoding is not required and the signal is directly transmitted over the medium connecting transmitter and receiver. Such signals for transmission are also referred to as 'lowpass signals'.

5.2.2 Passband Communication Systems

Unlike baseband communication systems, passband communication systems use carriers to modulate the message signals in such a way that the carrier frequency controls the translation of frequency spectrum of message signal. Further, using different carrier frequencies with certain difference, multiple modulated signals can be transmitted over a single line. Such a technique of transmission of multiple signals over a single line with different carrier frequencies is known as 'frequency division multiplexing'. The multiple signals transmitted over a single line can be separated using a bandpass filter centered at the carrier frequency of a particular modulated signal. So, such signals are also known as 'bandpass signals'. Also, the name 'passband' is also suggested based on the filter characteristics required to extract the required signal out of the received signal at the receiver.

Modulated signals multiplexed through frequency division multiplexing can also be transmitted over wireless media. Most wireless communication systems use passband signals for transmission to avoid interference. In this kind of communication systems, both analog and digital signals can be used as message signals. The message signals are transmitted using an appropriate modulation scheme. If the message signal is analog, any of the modulation schemes discussed in Chapter 3 can be used. The passband modulation schemes for digital message signals are discussed in this chapter in the sections to come.

Now, we discuss the pulse-based communication systems for analog signal transmission. The advantage of pulse based transmission is that it enables multiple message signal transmission over single line through time division multiplexing.

5.3 ANALOG MODULATION FOR PULSE-BASED SYSTEMS

Pulse-based signal transmission falls under the category of baseband communication, as the spectrum of such signals is concentrated around zero frequency. We categorize such modulation as analog modulation because the messages that we consider in this section are in the form of analog signals. We discuss this topic in the digital communication chapter because the carrier signal is a pulse train which is a digital signal. Basically, there are three types of pulse-based modulation schemes for analog messages: (1) pulse amplitude modulation (PAM), (2) pulse position modulation (PPM), and (3) pulse width modulation (PWM).

In case of modulation schemes with a sinusoidal carrier signal, discussed in Chapter 3, the parameters that can be modulated are amplitude, phase, and frequency. Similarly, the modulation type is based on the parameter that can be modulated in pulse-based communication systems.

To represent a digital carrier as a pulse train, let's define the pulse of duration t_p as

$$p(t)=\begin{cases}1, & |t|\le\frac{t_p}{2}\\ 0, & |t|>\frac{t_p}{2}\end{cases} \tag{5.1}$$

The pulse, $p(t)$, is repeated periodically to obtain the pulse train which can be used as carrier in pulse-based modulation schemes. Based on the above pulse definition, we represent the digital carrier signal pulse train as

$$c_d(t)=\sum_{n=-\infty}^{\infty}p(t-nT_s) \tag{5.2}$$

where T_s is the time after which a pulse repeats, i.e., pulse train has rate of $\frac{1}{T_s}$ pulses per second.

5.3.1 Pulse Amplitude Modulation

As the name suggests, pulse amplitude is changed according to instantaneous value of the message signal in PAM. In practice, PAM is the same as natural sampling that is discussed in the previous chapter. An example of PAM signalling has been shown in Figure 5.1. Mathematically, PAM can be represented as multiplication of the message signal with pulse train, i.e.,

$$S_{PA}(t) = m(t)c_d(t) \tag{5.3}$$

where $c_d(t)$ is the pulse train or the digital carrier signal. Although the mathematical representation of PAM looks like that of DSB-SC modulation scheme, the frequency spectrum of the message signal does not shift to a higher frequency band. The reason is that pulse train $p(t)$ is a periodic signal, and it can be represented as a Fourier series. Unlike the spectrum of PAM signal, $s_{PA}(t)$ can be represented in the same

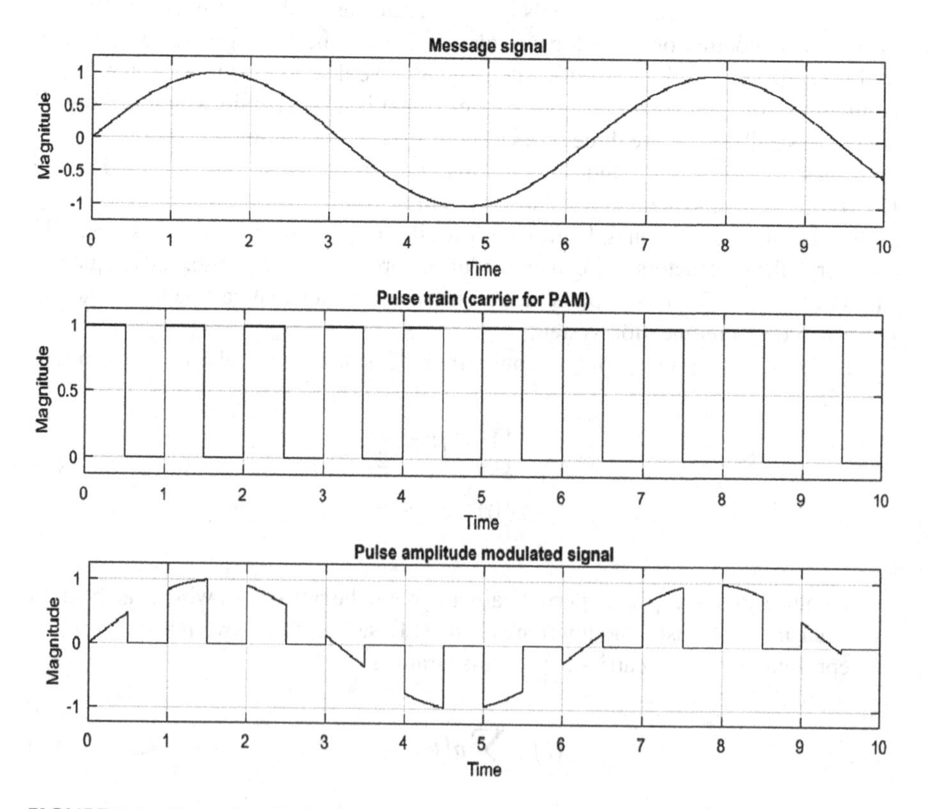

FIGURE 5.1 Example of pulse amplitude modulated signal.

manner as for a sampled signal. For this reason, the frequency of the pulse train used as a carrier for PAM shall be defined in accordance with the sampling theorem to avoid aliasing. This is also required for proper reconstruction of $m(t)$ at the receiver from a PAM signal.

Note that the bandwidth of a PAM signal, $S_{PA}(t)$, is infinite in accordance with the sampling process. However, since the pulses used are of finite width, the magnitude decreases as we go for higher harmonics of the pulse train frequency. So, the PAM signals can be represented with finite bandwidth considering significant power in the baseband frequency component. Accordingly, sufficient bandwidth of the transmission line is required for such a signal transmission.

Furthermore, a flat top sampled signal is also considered as a kind of PAM. It may be referred to as a flat top PAM.

5.3.2 Pulse Position Modulation

In pulse position modulation (PPM), the position of pulse is delayed or advanced as per the instantaneous value of the message signal. The PPM signal can be represented as a time shifted pulse where the shift is controlled by the instantaneous value of the message signal. It can be shown in mathematical representation as

$$S_{PP}(t) = c_d\left(t - k_p m(t)\right) \tag{5.4}$$

where $c_d(t)$ is the pulse train or the digital carrier signal and k_p is the constant that controls the time shift of pulses. In this type of representation, the pulses are shifted as follows:

- When the value of k_p is positive, the pulses are delayed for positive values of the message signal, and the pulses are advanced for negative values of the message signal.
- When the value of k_p is negative, the pulses are advanced for positive values of the message signal, and the pulses are delayed for negative values of the message signal.

An example of PPM signalling is shown in Figure 5.2. Furthermore, the PPM signal can be regarded as the phase modulation that was discussed in Chapter 3 (See expression (3.32)). The difference is that, in phase modulation, the carrier is a sinusoidal signal, while in PPM it is a pulse train. So, the frequency domain analysis would follow from the similar steps. But, the resulting spectrum of PPM falls in the baseband range. Hence, if multiple signals are to be transmitted over a single line, then time division multiplexing is a possible solution.

5.3.3 Pulse Width Modulation

As the name suggests, the width of the pulse is changed according to instantaneous value of the message signal in pulse width modulation (PWM). We know that time scaling of a time limited signal controls the width or duration of the signal. So, PWM

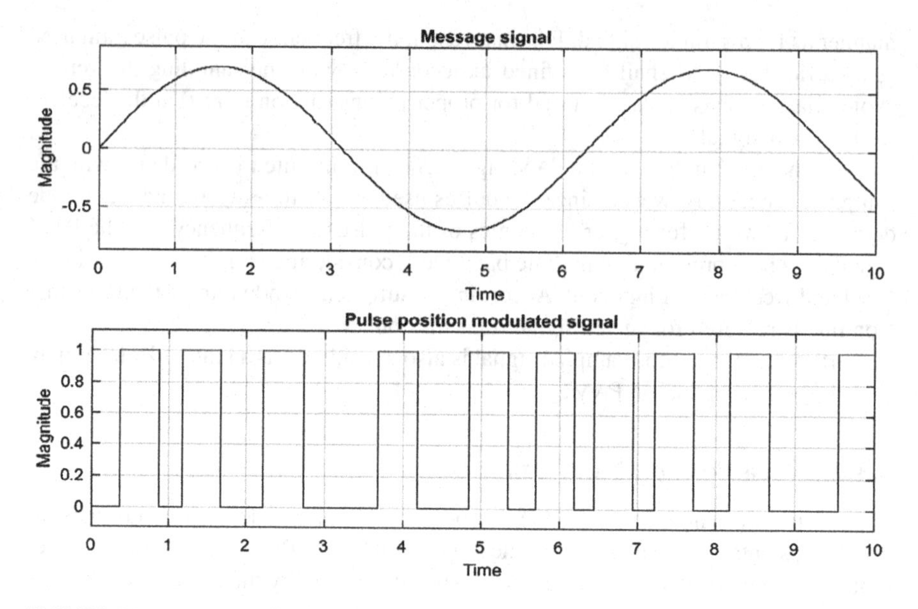

FIGURE 5.2 Example of pulse position modulated signal.

is considered as time scaling of the pulse train at respective instants of pulse occurrence. A PWM signal can be represented as

$$s_{PW}(t) = c_d\left(\frac{k_w}{m(t)}t\right) \tag{5.5}$$

where k_w is the constant that controls the width of pulses. An example of PWM signalling is shown in Figure 5.3. It can be observed from the figure that the duty cycle of each pulse is a function of instantaneous value of the message signal.

Further, the PWM signal can be regarded as the frequency modulation that was discussed in Chapter 3 (See expression (3.28)). However, PWM does not fundamentally represent a behavior similar to frequency modulation owing to the fact that only duty cycle of pulses changes in PWM and not the rate at which the pulses are repeated. So, the frequency domain analysis would follow from the similar steps. But, the resulting spectrum of PWM falls in the baseband range. Hence, if multiple signals are to be transmitted over a single line, then time division multiplexing is a possible solution.

5.4 CARRIER MODULATION FOR DIGITAL SYSTEMS

From the discussion in the previous section on pulse-based modulation schemes for analog message signals, we can establish an analogous pulse-based digital modulation scheme. In this section, we will discuss carrier modulation for digital communication systems. In such systems, the message signal is in digital form and the carrier signal is a continuous time sinusoidal signal. The usual impression about a digital

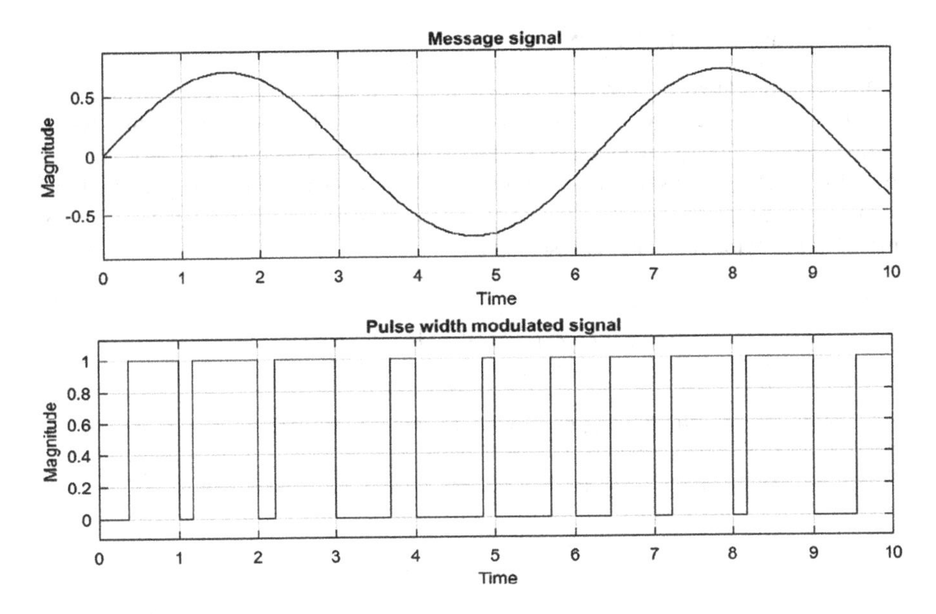

FIGURE 5.3 Example of pulse width modulated signal.

signal is that a signal is represented in binary form, i.e., in terms of bit '0' and bit '1'. However, we may consider M-ary representation of digital signals in this section. M-ary representation can be differentiated from binary representation through a simple description as follows:

- In binary form, we consider only two symbols. These symbols are represented by single bit.
- In M-ary form, we consider total M distinct symbols. Each of these symbols can be considered to represent $\log_2 M$ number of bits for its binary counterpart.

So, if $M = 4$ in an M-ary digital representation, we can consider each symbol as a combination of two bits, viz., '00', '01', '10' and '11', to represent all four symbols. However, the way of representing the symbol may vary.

In our subsequent discussions on carrier-based modulation schemes for digital communication systems, we will consider M-ary representation of digital signals. Hence, we will call such signals as M-ary signals and such modulation schemes as M-ary modulation schemes. It should be noted that binary signals are the special case of M-ary signals for $M = 2$.

Digital modulation schemes can be broadly classified into three types: (1) amplitude shift keying (ASK), (2) phase shift keying (PSK), and (3) frequency shift keying (FSK).

Further, each of these types of digital modulation schemes is directly related to its analog scheme of modulation. The main difference from analog modulation schemes is in the message signal.

5.4.1 Amplitude Shift Keying

Amplitude shift keying (ASK) can be defined in the same way as we defined amplitude modulation, i.e., the amplitude of the carrier signal is varied according to the instantaneous value of the message signal. Now, the difference in ASK as compared to AM is that the fixed set of possible values of message signal are being considered because the message signal is in digital form. In ASK, the modulated signal for arrival of mth symbol can be represented as

$$s_{m,ASK}(t) = A_m \cos(2\pi f_c t), \quad 0 \leq t < T \tag{5.6}$$

where A_m is the amplitude level to represent m^{th} symbol in M-ary representation and T is the symbol duration.

Usually, A_m is represented as

$$A_m = (2m - 1 - M)A \tag{5.7}$$

where $m \in \{1, 2, \ldots, M\}$ for M-ary message signal and A is the arbitrary amplitude which can be considered as the carrier signal amplitude without loss of generality.

Unlike other schemes, we represent m^{th} symbol as a modulated signal. One can get modulated symbol representations for all symbols considering the values of m from 1 to M. So, depending on the order of arrival of symbols, the modulated signal can be obtained by selecting modulated symbol corresponding to the incoming symbol sequence in the message signal.

5.4.2 Phase Shift Keying

Phase shift keying (PSK) modulation is the digital modulation scheme corresponding to phase modulation in the analog communication systems. In PSK, the phase of the carrier signal varies according to the instantaneous value of the incoming message symbol. We realize the fact that the message signal is in digital form. So, the PSK signal would comprise of a finite number of discrete phases in the modulated signal. In PSK, the modulated signal for arrival of mth symbol can be represented as

$$s_{m,PSK}(t) = A\cos(2\pi f_c t + \theta_m), \quad 0 \leq t < T \tag{5.8}$$

where θ_m is the phase shift corresponding to m^{th} symbol in M-ary representation, and T is the symbol duration. The above representation can be further simplified to represent the modulated signal for arrival of mth symbol as

$$s_{m,PSK}(t) = A\cos(2\pi f_c t)\cos(\theta_m) - A\sin(2\pi f_c t)\sin(\theta_m), \quad 0 \leq t < T \tag{5.9}$$

From the above representation, we note the following points:

1. PSK is represented as a summation of two ASK signals.
2. Carriers used for the two ASKs are orthogonal to each other and sinusoidal signals.
3. Carriers are $A\cos(2\pi f_c t)$ and $-A\sin(2\pi f_c t)$.
4. Message signals corresponding to these carriers are $\cos(\theta_m)$ and $\sin(\theta_m)$, respectively.

In addition, the phase can take one of the possible values in the range $(0, 2\pi)$. So, the entire range of interval 2π is divided by M to obtain equally spaced representation in terms of phase angle. Thus, the representation of phase shift corresponding to m^{th} symbol, θ_m is given as

$$\theta_m = \frac{2\pi}{M} m \tag{5.10}$$

where m takes values from any of the following possibilities:

- $m \in \{1, 2, \ldots, M\}$ for M-ary message signal. In this case, $\theta_M = 2\pi$ which is the same as $\theta_M = 0$
- $m \in \{0, 1, \ldots; M-1\}$ for M-ary message signal. In this case, $\theta_0 = 0$ which is the same as $\theta_0 = 2\pi$

Both of the above representations obtain the same values of θ_m with one representing the other with a rotation of $\frac{2\pi}{M}$.

It should be noted that the symbol energy for any value of m is the same in this case. So, PSK is also known as a constant energy modulation scheme.

Unlike other schemes, we represent m^{th} symbol as a modulated signal. One can get modulated symbol representations for all symbols considering the values of m from 1 to M. So, depending on the order of arrival of symbols, the modulated signal can be obtained by selecting a modulated symbol corresponding to the incoming symbol sequence in the message signal.

5.4.3 Frequency Shift Keying

As the name suggests, the frequency of the carrier changes according to the incoming symbol of the message signal sequence in frequency shift keying (FSK). As noted for ASK and PSK, FSK is the digital modulation scheme corresponding to frequency modulation. But, since the message signal takes only discrete values and is in digital form, the FSK modulated signal will have carriers with different frequency coming from a finite set of carrier frequency values. In FSK, the modulated signal for arrival of mth symbol can be represented as

$$s_{m,FSK}(t) = A\cos\left[2\pi\left(f_c + m\Delta f\right)t\right], \quad 0 \le t < T \tag{5.11}$$

where $m\Delta f$ is the carrier frequency deviation corresponding to m^{th} symbol in M-ary representation and T is the symbol duration and $m \in \{1,2,\ldots,M\}$.
Unlike the other modulation schemes discussed earlier in this book, the set of frequencies to be used for FSK have to be orthogonal to each other, i.e., any two carrier frequencies with deviation $m\Delta f$ and $n\Delta f$ from the set of possible frequencies would satisfy the following condition:

$$\int_0^{-T} \cos\left[2\pi\left(f_c + m\Delta f\right)t\right]\cos\left[2\pi\left(f_c + n\Delta f\right)t\right]dt = 0 \tag{5.12}$$

Using the above condition for orthogonality of carrier frequencies, we get the required condition to obtain an orthogonal set of carrier frequencies on the carrier frequency deviation as

$$\Delta f = \frac{k}{2T} \tag{5.13}$$

where $k = 0$ is an integer. Using the above expression, we obtain the minimum carrier frequency deviation for $k = 1$ which can be given as

$$\Delta f_{mim} = \frac{1}{2T} \tag{5.14}$$

So, a symbol is mapped to a carrier frequency in FSK. An FSK signal is transmitted as a sequence of orthogonal frequency carrier frequencies according to the incoming message signal symbols.

5.5 CONSTELLATION DIAGRAM AND EFFECT OF NOISE ON DIGITAL MODULATION SCHEMES

From the discussion in the previous section on ASK, PSK, and FSK, we note the following:

1. ASK can be realized using a single carrier.
2. PSK can be represented using two orthogonal carriers: $\cos(2\pi f_c t)$ and $\sin(2\pi f_c t)$.
3. FSK can be represented using M orthogonal carriers for an M-ary FSK system represented as $\cos\,[2\pi(f_c + m\Delta f)\,t]$ with suitable values of Δf.

Therefore, using the concept of Gram-Schmidt orthogonalization for vector representation of signals, we can represent ASK, PSK, and FSK as a single dimension, two-dimensional and M dimensional vectors, respectively. Interested readers may refer to appendix A of [1] for a detailed explanation on vector representation of signals and Gram-Schmidt orthogonalization.

Let a signal $s_m(t)$ be denoted by vector s_m in its vector representation. As per the usual vector representation, a vector has a magnitude and a direction. The direction is specified by a number of unit vectors. The number of unit vectors is the same as the dimension of that vector. In terms of linear algebra, we refer to these unit vectors as an orthonormal basis.

Some of the properties of vectors contained in an orthonormal basis are as follows:

1. Each vector of an orthonormal basis has magnitude one, i.e., the length of each vector in an orthonormal basis is one.
2. As per the property of orthonormality, these vectors are orthogonal to each other.

Let us consider the case of a 3D vector, a. The most common representation for such a vector is

$$\mathrm{a} = x\mathrm{a}_{\mathrm{x}} + y\mathrm{a}_{\mathrm{y}} + z\mathrm{a}_{\mathrm{z}} \tag{5.15}$$

where

- a_x, a_y, and a_z are unit vectors representing the orthonormal basis for the vector a.
- x, y, and z are the vector components in the directions a_x, a_y, and a_z respectively.
- x, y, and z represent the components of orthogonal decomposition of vector a.
- The directions of unit vectors, a_x, a_y, and a_z are usually referred to as the directions X, Y, and Z in accordance with a 3D cartesian coordinate system.

As a generalized case, an N dimensional vector, v, is represented as

$$\mathrm{v} = \sum_{i=1}^{N} v_i \mathrm{a}_{\mathrm{i}} \tag{5.16}$$

where

- v_i is the ith component of vector v.
- N is the dimension of vector v.
- a_i is the ith unit vector in the orthonormal basis of vector v.

Similar to the representation of vector in terms of summation of its orthogonal decomposed components, the vector representation of a signal can be represented in the form

$$s_m(t) = \sum_{i=1}^{N} s_{m,i}\varphi_i(t) \tag{5.17}$$

where

- $s_{m,i}$ is the ith component of vector s_m.
- N is the dimension of vector s_m.
- $\varphi_i(t)$ is the ith basis function in the orthonormal basis for representation of $s_m(t)$ in its vector form.

Further, it should be noted that the energy of the signal represented using the above expression (5.17) can be given by the squared length of the vector that represents the signal, i.e., in the case under consideration,

$$\begin{aligned}\int_{-\infty}^{-\infty} \left|s_m(t)\right|^2 dt &= \sum_{i=0}^{N} \left|s_{m,i}\right|^2 \\ &= \left|s_m\right|^2\end{aligned} \tag{5.18}$$

Properties of basis function
Some of the properties of basis functions for vector representation are as follows:

1. Each $\varphi_i(t)$ is a unit energy signal, i.e.,

$$\int_{-\infty}^{-\infty} \left|\varphi_i(t)\right| dt = 1 \tag{5.19}$$

2. $\varphi_i(t)$ and $\varphi_i(t)$ are orthonormal for all possible values satisfying $i/= j$, i.e.,

$$\int_{-\infty}^{-\infty} \mathrm{I}\varphi_i(t)\varphi_j(t) dt = 0, i = j \tag{5.20}$$

Constellation diagram
A constellation diagram of a set of signals is obtained as graphical representation of the signals in their vector form. This is usually referred to in the context of set of signals in a particular digital modulation scheme.

Based on the dimensionality of a modulation scheme, we represent the constellation diagram using vector representation of the signals. Note that for an M-ary modulation scheme, we get M vectors in the vector representation and M points in the corresponding constellation diagram. Now, we discuss the vector representation of each modulation scheme that we discussed in the previous section.

5.5.1 Constellation Diagram for ASK

In ASK, we represent the modulation symbol as

$$s_{m,ASK}(t) = A_m \cos(2\pi f_c t), \quad 0 \le t < T \tag{5.21}$$

where $A_m = (2m - 1 - M)\,A$. Comparing the above expression with Equation (5.17), we obtain the following:

$$N = 1$$

$$\varphi_1(t) = \sqrt{\frac{-2}{T}} \cos(2\pi f_c t)$$

$$s_{m,ASK,1} = \sqrt{\frac{-T}{2}} (2m - 1 - M) A$$

Considering $A = \sqrt{\frac{-2}{T}}$ the mth symbol modulated with ASK scheme can be represented in the form of vector as

$$s_{mASK} = 2m - 1 - M \tag{5.22}$$

Based on the vector representation of ASK, we note the following:

1. ASK is a one-dimensional constellation.
2. The distance between any two constellation points can be given by

$$d_{mn} = | s_{mASK} - s_{nASK} | = 2\,| m - n |.$$

3. The energy of the ASK modulated signal on arrival of mth symbol can be given by

$$E_{m,ASK} = |s_m|^2 = (2m - 1 - M)^2 \tag{5.23}$$

A constellation diagram of ASK modulation scheme is shown in Figure 5.4 for $M = 4$ and $A = \sqrt{\frac{2}{T}}$ in Figure 5.4(a) and $M = 8$ and $A = \sqrt{\frac{2}{T}}$ in Figure 5.4(b). This scheme is also referred to as M-ASK, i.e., 8-ASK scheme for the one with an $M = 8$ point constellation. In this constellation, each point represents a symbol of length $\log_2 8 = 3$ bits. The assignment of symbols may be based on a gray code so that the 3-bit codes corresponding to adjacent symbols differ only in single bit. It should be noted from the figure that as the number of constellation points, M, increases, the additional constellation points are added which would be farther from the origin. So, such additional points increase the average energy per symbol in the ASK constellation.

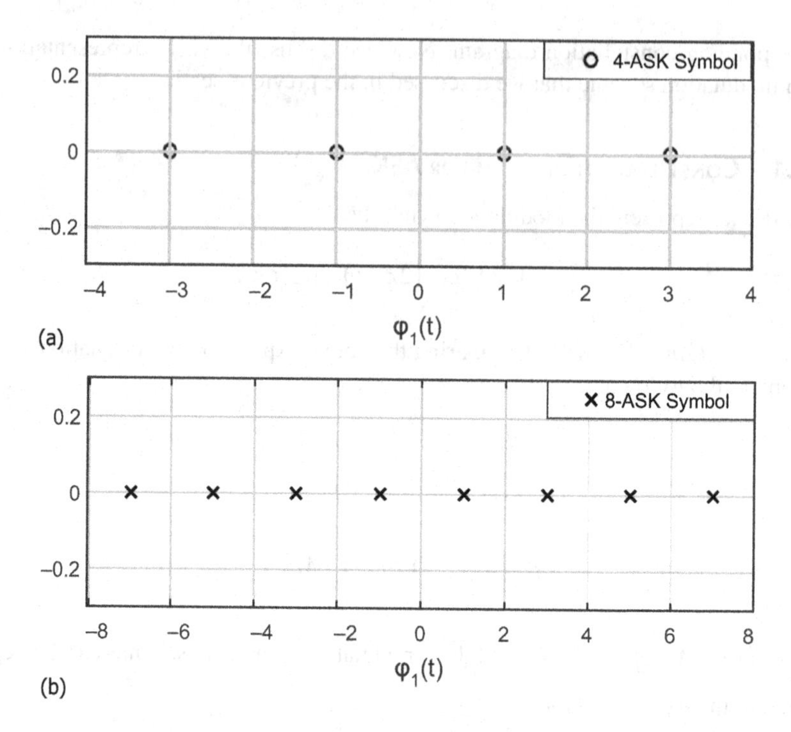

FIGURE 5.4. Constellation diagram of ASK modulation scheme: (a) 4-ASK for $M = 4$ and $A = \sqrt{\frac{2}{T}}$, (b) 8-ASK for $M = 8$ and $A = \sqrt{\frac{2}{T}}$

5.5.2 Constellation Diagram for PSK

In a PSK modulation scheme, the modulated signal for arrival of mth symbol can be represented as

$$s_{m,PSK}(t) = A\cos(2\pi f_c t)\cos(\theta_m) - A\sin(2\pi f_c t)\sin(\theta_m), 0 \le t < T \quad (5.24)$$

where $\theta_m = \frac{2\pi}{M}m$. Comparing the above expression with Equation (5.17), we obtain the following:

$$N = 2$$

$$\varphi_1(t) = \sqrt{\frac{-2}{T}}\cos(2\pi f_c t)$$

$$\varphi_2(t) = \sqrt{-\frac{-2}{T}}\cos(2\pi f_c t)$$

$$s_{m,PSK,1} = \sqrt{\frac{-T}{2}} A\cos(\theta_m)$$

$$s_{m,PSK,2} = \sqrt{\frac{-T}{2}} A\sin(\theta_m)$$

Considering $A = \sqrt{\frac{2}{T}}$, the mth symbol modulated with PSK scheme can be represented in the form of a vector as

$$\mathrm{s}_{\mathrm{m}_{PSK}} = \left[\cos\left(\frac{2\pi}{M}m\right), \sin\left(\frac{2\pi}{M}m\right)\right] \quad (5.25)$$

Based on the vector representation of PSK, we note the following:

1. PSK is a 2D constellation with unit energy sine and cosine signals as an orthonormal basis.
2. The distance between any two constellation points can be given by

$$d_{mn} = \left|\mathrm{s}_{\mathrm{m}_{PSK}} - \mathrm{s}_{\mathrm{n}_{PSK}}\right|.$$

3. The energy of the PSK modulated signal on arrival of mth symbol can be given by

$$E_{m,PSK} = |\mathrm{s}_{\mathrm{m}}|^2 = \cos(\theta_m)^2 + \sin(\theta_m)^2 = 1, \quad (5.26)$$

i.e., all the constellation points carry equal energy. Moreover, the energy of each symbol can be controlled by setting suitable value of A, but this would still result in all the points with equal energy in a PSK modulation scheme.

A constellation diagram of 8-PSK modulation scheme is shown in Figure 5.4 for $M = 8$ and $A = \sqrt{\frac{2}{T}}$. It can be observed that all the constellation points are on the unit radius circle, i.e., they are equidistant from the origin. This justifies the fact that each symbol carries the same energy in PSK modulation scheme.

5.5.3 Constellation Diagram for FSK

In a FSK modulation scheme, the modulated signal for arrival of mth symbol can be represented as

$$s_{m,FSK}(t) = A\cos\left[2\pi(f_c + m\Delta f)t\right], 0 \le t < T \quad (5.27)$$

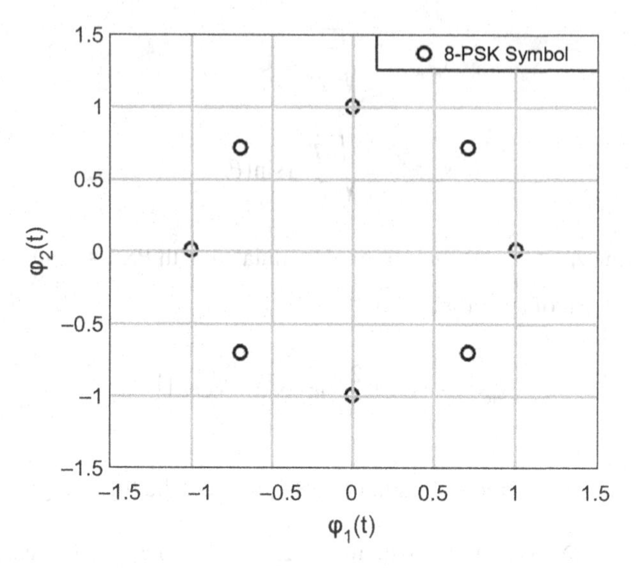

FIGURE 5.5 Constellation diagram of 8-PSK modulation scheme with $M = 8$ and $A = \sqrt{\frac{2}{T}}$.

In our discussion on FSK systems in the previous section, we established that the symbols represented by $s_{m,FSK}(t)$ are orthogonal to each other. So, we can represent the modulated signal for arrival of mth symbol alternatively as

$$s_{m,FSK}(t) = \sum_{i=1}^{M} a_i A \cos\left[2\pi\left(f_c + i\Delta f\right)t\right], 0 \le t < T \tag{5.28}$$

where $a_i = 1$ for $i = m$ and $a_i = 0$ for all the values of i other than $i = m$. This representation is in accordance with the generalized representation of signals in terms of their vector components and basis functions. Comparing the above expression with Equation (5.17), we obtain the following:

$$N = M$$

$$\varphi_1(t) = \sqrt{\frac{-2}{T}} \cos\left(2\pi f_c t + \Delta f\right)$$

$$\varphi_2(t) = \sqrt{\frac{-2}{T}} \cos\left(2\pi f_c t + 2\Delta f\right)$$

$$\vdots$$

$$\varphi_M(t) = \sqrt{\frac{-2}{T}} \cos\left(2\pi f_c t + M\Delta f\right)$$

$$s_{m,FSF,k} = \begin{cases} 0\sqrt{\frac{T}{2}}A & \text{for} \quad k = m \\ & \text{for} \quad k/ = m \end{cases}$$

Considering $A = \sqrt{\frac{2}{T}}$, the mth symbol modulated with FSK scheme can be represented in the form of vector as

$$\mathrm{s}_{\mathrm{m}_{FSK}} = \left[0,0,\cdots,0, \underbrace{1}_{m^{th}\,position} ,0,\cdots,0 \right] \tag{5.29}$$

Based on the vector representation of FSK, we note the following:

1. FSK is an M dimensional constellation. Vector representation of a symbol in FSK contains only one non-zero component as observed in expression (5.29).
2. The distance between any two constellation points is the same and it is given by $d_{mn} = \left| \mathrm{s}_{\mathrm{m}_{FSK}} - \mathrm{s}_{\mathrm{n}_{FSK}} \right| = \sqrt{2}$ for $m \neq n$.
3. The energy of the FSK modulated signal on arrival of mth symbol can be given by

$$E_{m,FSK} = \left| \mathrm{s}_{\mathrm{m}} \right|^2 = \sum_{i=1}^{M} s_{m,FSK,i} = 1, \tag{5.30}$$

i.e., all the constellation points carry equal energy. Moreover, the energy of each symbol can be controlled by setting suitable value of A, but this would still result in all the points with equal energy in a FSK modulation scheme.

Being an M dimensional modulation scheme, it is difficult to represent the constellation diagrams of M-FSK modulation scheme for $M > 2$. A constellation diagram of 2-FSK modulation scheme is shown in Figure 5.6 for $M = 2$ and $A = \sqrt{\frac{2}{T}}$. All the higher order FSK modulation cannot be observed graphically. However, M dimensional vectors are used to represent such symbols.

5.5.4 Effect of Noise on Constellation Diagram

While propagating through a wired or wireless media, a transmitted signal does not reach the receiver in its original form. In addition to certain effects due to channel characteristics, the signal is affected by noise. The most common type of noise affecting the communication signal is considered to be additive white Gaussian noise (AWGN). Certain characteristics of AWGN are as follows:

- It is additive in nature, i.e., the signal, s, becomes $s + n$ after addition of noise n.
- Noise is a random process. The probability distribution function of the samples collected at uniform interval is Gaussian distributed. For reference, the Gaussian PDF is given by

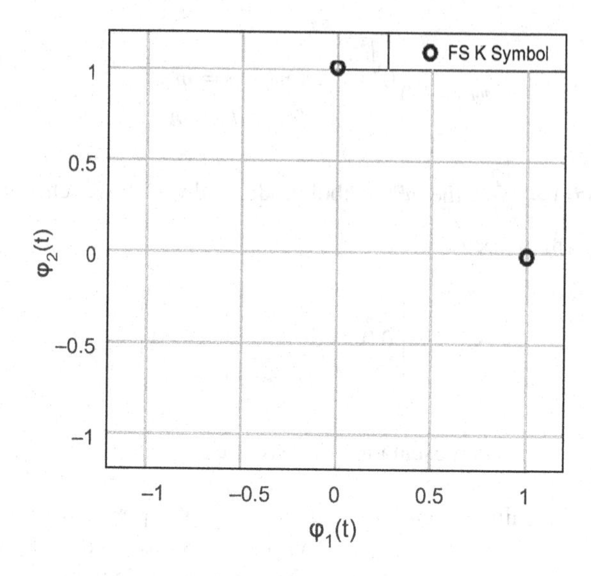

FIGURE 5.6. Constellation diagram of 2-FSK modulation scheme with $M = 2$ and $A = \sqrt{\frac{2}{T}}$

$$p_x(x) = \frac{1}{\sqrt{2\pi\sigma_x^2}} e^{-\frac{(x-\mu_x)^2}{2_x\sigma^2}} \tag{5.31}$$

where μ_x is the mean of random variable X and σ_x^2 is the variance of random variable X.

- The mean of such noise samples is zero. So, the power of such noise process is the same as its variance.
- The autocorrelation function of AWGN is an impulse function.
- The power spectral density, which is Fourier transform of autocorrelation function, has uniform magnitude for all the frequency. This is the reason it is called 'white noise'.

The noise is also added by the electronic components used in the signal processing circuitry. Such noise sources are independent noise generators and large in number; therefore, it is justified to consider the noise to be Gaussian distributed as in case of AWGN. This is in accordance with the central limit theorem [2] which has wide applications in signal processing and communication theory development where multiple noise sources or multiple random processes/variables are involved [3,4,5].

Considering the noise being added to the received signal, the signal representations for various modulation schemes no longer remain the same as we did for ASK in expression (5.21), for PSK in expression (5.24), and for FSK in expression (5.28). All these signals will have some noise components added to them. The noise would disturb the position of a constellation point. It should be noted that the dimension of noise would be greater than the signal dimension in general because the symbols represented in a particular modulation scheme come from a fix set of possible symbols. However, the noise is generated from various sources. But, it would be sufficient to consider only those components of noise which belong to projection on the basis functions of the modulation scheme under consideration [6].

To analyze the effect of noise on the signal constellation, consider the case of PSK modulated signals. The modulated signal is represented as

$$s_{m,PSK}(t) = A\cos(2\pi f_c t)\cos(\theta_m) - A\sin(2\pi f_c t)\sin(\theta_m), \quad 0 \le t < T \tag{5.32}$$

Such that we obtain two basis functions for representation in the vector form. The basis functions are

$$\varphi_1(t) = \sqrt{\frac{-2}{T}}\cos(2\pi f_c t)$$

and

$$\varphi_2(t) = \sqrt{-\frac{-2}{T}}\sin(2\pi f_c t).$$

Considering $A = \sqrt{\frac{2}{T}}$, PSK scheme can be represented in the form of vector as

$$s_{mPSK} = \left[\cos\left(\frac{2\pi}{M}m\right), \sin\left(\frac{2\pi}{M}m\right)\right] \tag{5.33}$$

Now, consider the received signal in some symbol duration is

$$r(t) = s_{PSK}(t) + n(t) \tag{5.34}$$

where $n(t)$ is the AWGN process. To represent $n(t)$ in vector form, we get the following:

$$n_1 = \int_0^{-T} n(t)\varphi_1(t)dt$$

$$n_2 = \int_0^{-T} n(t)\varphi_2(t)dt$$

where n_1 is the component in direction of $\varphi_1(t)$ and n_2 is the component in direction of $\varphi_2(t)$, i.e., the correlation of noise process with the corresponding basis function of PSK signal representation. There can be more vector components of the noise process, $n(t)$. Those vector components when added to the PSK signal would not affect it. So, the received signal in vector form using the transmit signal vector form as in Equation (5.35) can be given as

$$\mathrm{r} = \left[\cos\left(\frac{2\pi}{M}m\right) + n_1, \sin\left(\frac{2\pi}{M}m\right) + n_2\right] \tag{5.35}$$

As it can be seen from the above expression, the location of received signal vector differs from the transmitted signal vector. So, the constellation of received signals would be different from the constellation of transmitted signals. A constellation diagram to demonstrate the effect of AWGN on the signal constellation of PSK modulation scheme is shown in Figure 5.7 with signal to noise ratio (SNR) of 20 dB, i.e., the signal power is 100 times the noise power.

It is also interesting to see the effect of noise with different strengths. Given that we keep the transmit power of the signal the same and compare the constellation diagram under different noise powers, this would change the SNR values. Figure 5.8 shows received signal constellation points with SNR 10 dB and SNR 20 dB. It can be observed that the higher the SNR value, the lower would be the displacement of received constellation point from the transmitted constellation point. The received constellation points are more concentrated around the actual position of constellation point under less noisy conditions, i.e., higher SNR values. The constellation diagram of received symbols is important, as it tells about the performance of system in terms of proportion of bits/symbols received in error. Compare the received symbol constellation for 20 dB SNR shown in Figure 5.7 and the received symbol constellation for 10 dB SNR shown in Figure 5.8. The received constellation points can be

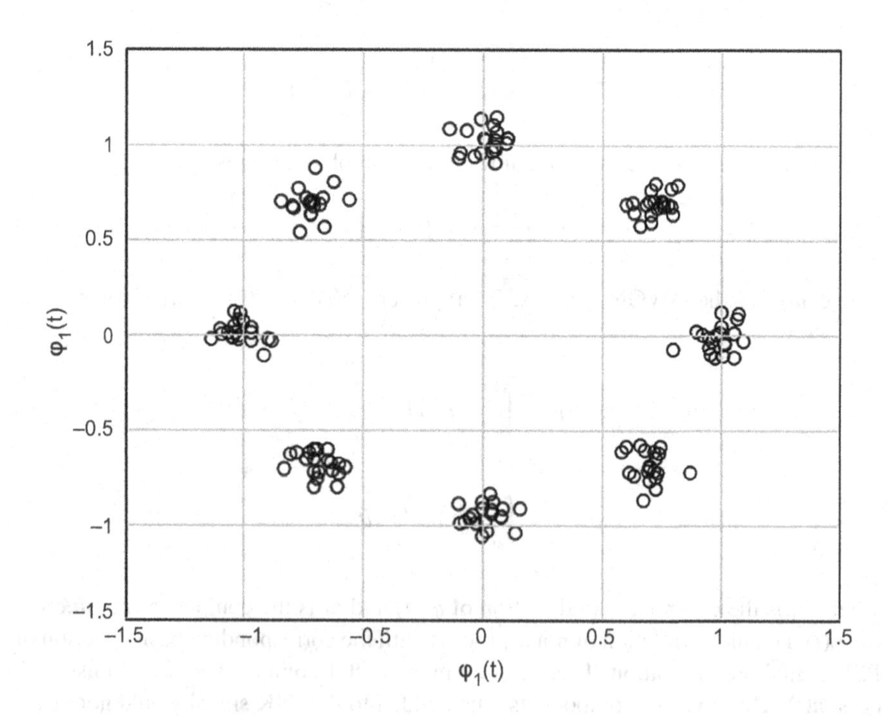

FIGURE 5.7 Constellation diagram of PSK modulation scheme under the effect of AWGN with an SNR of 20 dB.

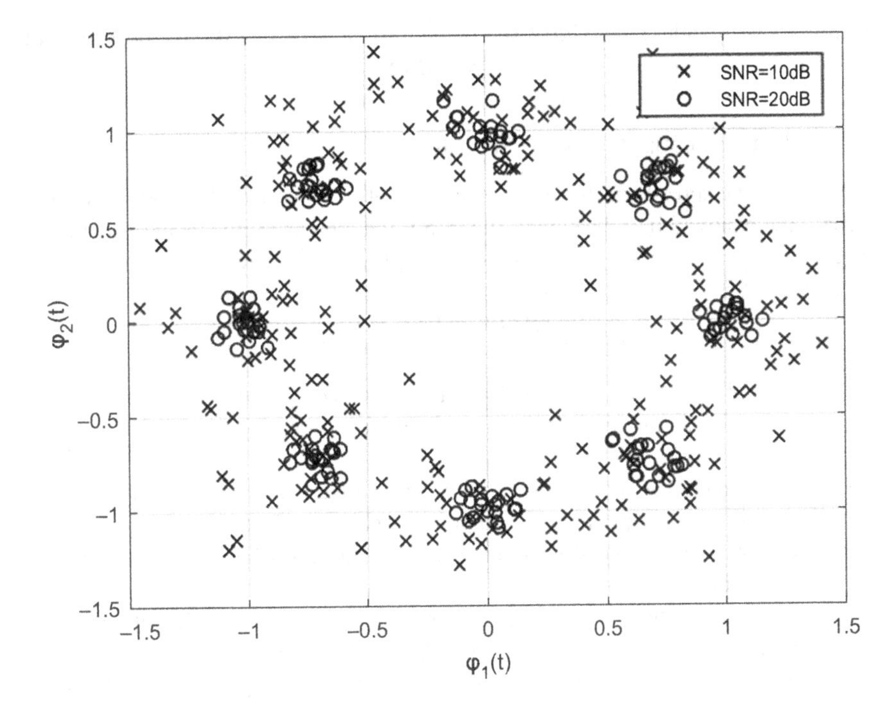

FIGURE 5.8 Constellation diagram of PSK modulation scheme under the effect of AWGN with an SNR of 10 dB and 20dB to observe the effect of different noise power on the constellation diagram.

classified into one of the transmit constellation points with almost sure decision in case of 20 dB SNR, while this surety of the classification would be less in case of received constellation points for 10 dB SNR. This surety has direct impact on the errors that we make in mapping the received constellation points to the transmitted constellation points.

5.6 SPREAD SPECTRUM MODULATION

Spread spectrum modulation is another kind of modulation in which a different principle is used. Usually, all the communication systems are designed in a way to use minimum bandwidth to transmit information at a maximum rate. However, the spectrum of signal is spread to much wider bandwidth than the signal bandwidth in spread spectrum modulation scheme. The bandwidth enhancement is done by adding a lot of redundant data in spread spectrum transmission. The redundant data transmission adds a level of security in the signal transmission. There are two ways in which the spectrum of the signal can be spread to much wider bandwidth than the required minimum bandwidth for signal transmission: (1) direct sequence spread spectrum (DS-SS), and (2) frequency hopping spread spectrum (FH-SS).

The code division multiple access (CDMA) standard of mobile communication systems is based on spread spectrum transmission technology. Furthermore, the same technology with wider bandwidth used in the third generation (3G) wireless technology with the name wideband CDMA (WCDMA).

5.6.1 Direct Sequence Spread Spectrum

Direct sequence spread spectrum uses the concept of redundant data transmission to increase the bandwidth of the signal. In DS-SS, the message signal is multiplied with a high rate baseband signal. A DS-SS signal, $s_d(t)$, can be represented as

$$s_d(t) = m(t)c(t) \tag{5.36}$$

where $m(t)$ is the message signal and $c(t)$ is the spreading signal. The properties of $m(t)$ and $c(t)$ and the relationship between them are summarized in brief below.

- Both the message signal and the spreading signal are in binary form.
- These signal are represented with polar signalling scheme of line encoding.
- The bit rate of the spreading signal is integer multiple of the bit rate of the message signal.
- One bit of the spreading signal is referred to as a 'chip', and the rate of arrival of chip in the spreading signal is referred to as a 'chip rate'.
- The ratio of bit rate of spreading signal to that of message signal is referred to as a 'spreading factor'. The spreading factor is the factor by which the bandwidth of message signal is spread.
- Spreading signal is usually a pseudo-random sequence. Such pseudo-random sequences can be generated using shift registers with suitable feedback mechanism.
- Using orthogonal pseudo-random sequences for multiple users, the DS-SS signals can be multiplexed.

5.6.2 Frequency Hopping Spread Spectrum

Similar to DS-SS, spreading of the spectrum of a message is realized using a high rate pseudo-random sequence. In FH-SS, the carrier frequency is switched as per the pseudo-random allotment. There are multiple carrier frequency changes within a bit duration. Furthermore, the duration of each carrier frequency is known as 'chip duration' in FH-SS. So, there are two ways to spread the spectrum in FH-SS:

1. having carrier frequencies far away from one another, and
2. having a high chip rate, i.e., frequent carrier frequency changes.

The FH-SS process can be understood better as follows:

- A channel is subdivided into subchannels such that each subchannel is orthogonal to each other in frequency domain. In other words, the subchannels have a non-overlapping frequency spectrum.

- As a carrier frequency is associated with a channel, a subcarrier frequency is associated with a subchannel. Let it be the center frequency of the subchannel.
- A bit duration of the message signal is divided into time slots. Each time slot duration is referred to as 'chip duration'.
- The number of time slots per bit duration and the number of subchannels are kept equal in general.
- A subchannel is assigned to every time slot of a bit duration at random, i.e., using pseudo-random fashion.
- A number of carrier frequency transition occurs within a bit duration.
- The total bandwidth of the signal is proportional to the number of time slots per bit duration and the spacing between the frequency of subcarriers.
- Multiple signals can be transmitted by assignment of a sequence of subchannels such that there is no overlap among subchannels in a particular time slot of a bit duration.

The same sequence of subcarrier frequencies needs to be generated in order to detect the symbol. A typical subchannel assignment strategy for an FH-SS system is shown in Figure 5.9. FH-SS technique is used in Bluetooth technology. Some of the advantages of FH-SS transmission technology are discussed below.

1. **Immunity to narrowband fading**
   The channels used for transmission in spread spectrum technique are inherently wideband channels. In FH-SS, the wide-band signal transmission consists of multiple narrowband subcarrier frequency subchannels. So, if one subchannel undergoes deep fade during the communication, the signal can be recovered/detected using other subchannel signals which may not be in deep fade condition at that moment. So, narrowband fading does not affect the performance of the system severely.

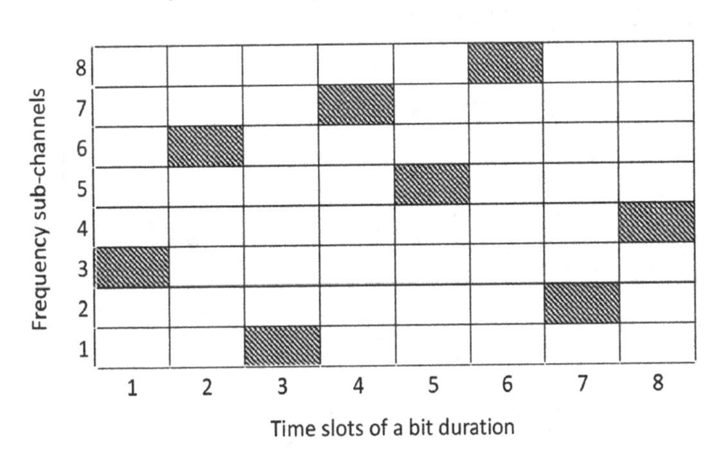

FIGURE 5.9 A typical subchannel assignment strategy for an FH-SS system considering 8 subchannels and 8 time slots per bit duration.

2. **Immunity to jamming**
   Mobile/wireless signal jammer uses noise signal as interference for the purpose of jamming the signals. Usually, these signals are of narrow band to have sufficient signal power per unit bandwidth for jamming. Because FH-SS signals are wideband signals containing multiple carriers carrying the same information bit, they would not get affected much. As explained in case of narrowband fading, the FHSS signal, if affected by a jamming signal at some subchannels, can be recovered using other subchannel signals which are not affected by the narrowband jamming signal at that moment.

5.6.3 Code Division Multiple Access

Code division multiple access (CDMA) is a technique to allow multiple users to share the same set of resources. The crucial resources in wireless communication systems are bandwidth and time-duration of channel availability.

The above resources are shared among users using one or the other technique of multiple access to have optimum use of resources while maintaining quality of service to each user.

Multiple users can be assigned time-frequency resources in such a way that the signals from different users do not interfere with each other. We have already discussed two techniques that allow multiple users to share the same set of resources. They are summarized below in brief:

1. Time division multiple access (TDMA)
   a. The time axis is divided into multiple slots in TDMA (Figure 5.10). Each user is assigned one or more slots based on traffic and availability of time slots per user. In TDMA, the effective time available for a user to transmit and receive signals is based on the number of slots assigned.
   b. Orthogonality of signals from multiple users is obtained in time domain.
2. Frequency division multiple access (FDMA)

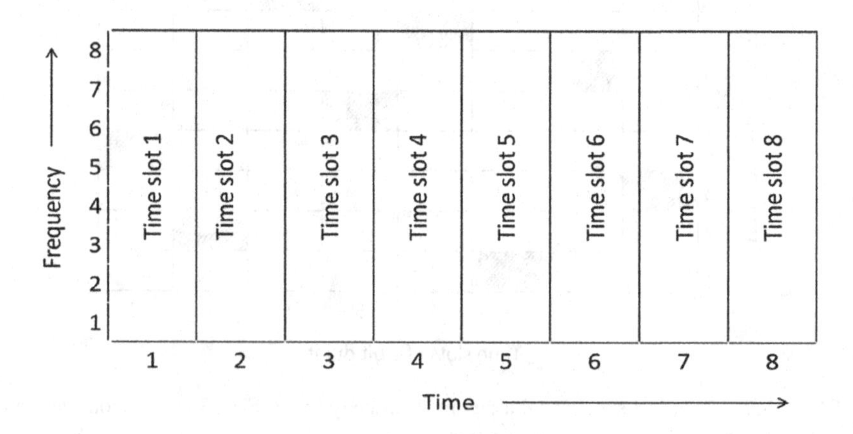

FIGURE 5.10 Time-frequency sharing in time division multiple access scheme.

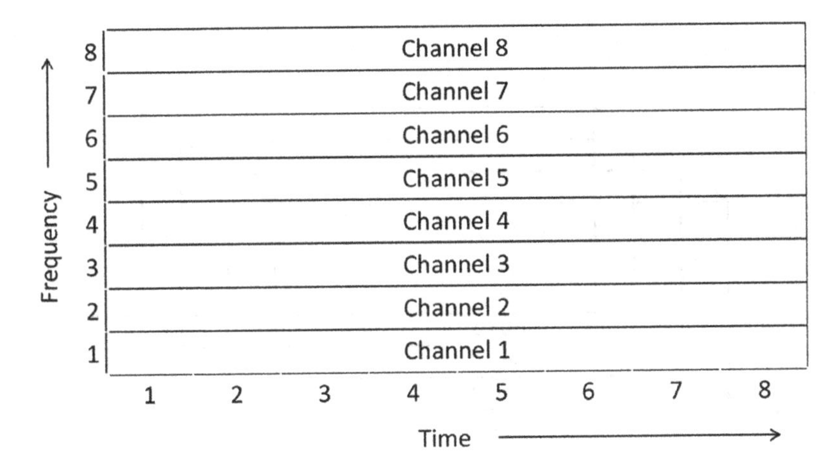

FIGURE 5.11 Time-frequency sharing in frequency division multiple access scheme.

a. The frequency band is divided into narrow bands called 'channels' in FDMA (Figure 5.11). A user is assigned a channel for the duration of communication. In FDMA, the throughput is limited by the channel bandwidth.
b. Orthogonality of signals from multiple users is obtained in frequency domain through non-overlapping channels.

On the other hand, CDMA allows access of the same set of resources to multiple users simultaneously. Spread spectrum technique is the backbone for CDMA. The different characteristics of CDMA as compared to TDMA and FDMA are as follows:

1. All users use complete frequency band all the time, i.e., neither time slots are created nor the frequency band is divided into narrowband channels.
2. CDMA adds one more dimension to maintain orthogonality among the signals from multiple users.
3. Signal orthogonality among the signals transmitted by multiple users is achieved using orthogonal spreading sequences, say, orthogonal pseudo-random spreading sequences commonly known as ''orthogonal codes'.
4. In DS-SS, orthogonal pseudo-random sequences are used for multiple users as spreading sequences.
5. In FH-SS, the orthogonality is achieved through pseudo-random channel allocation to multiple users in such a way that no two users are allotted same subchannel in the same time slot of a bit duration.

Resource sharing among users using orthogonal codes for CDMA has been shown in Figure 5.12. where Code 1, Code 2, Code 3, etc. denote the orthogonal spreading sequences. The orthogonal spreading sequences in DS-SS allows multiple users to transmit their signals simultaneously using the same frequency, i.e., in the same channel without any interference among their signals. While, in FH-SS the

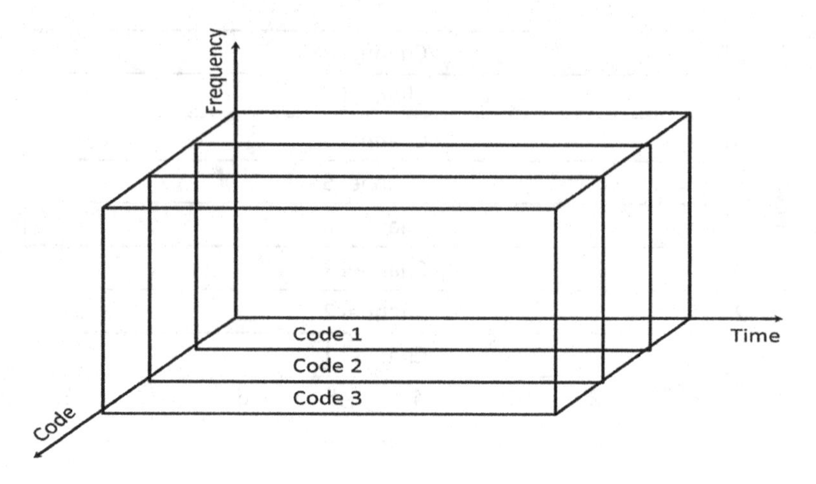

FIGURE 5.12 Resource sharing in code division multiple access scheme

subchannel allocation among the users is such that no two users are assigned same subchannel during a chip time.

Example of CDMA through FH-SS

An example of subchannel assignment strategy for FH-SS based CDMA system is shown in Figure 5.13. The subchannel assigned to different users in different chip durations is shown with different fill patterns as indicated in the figure.

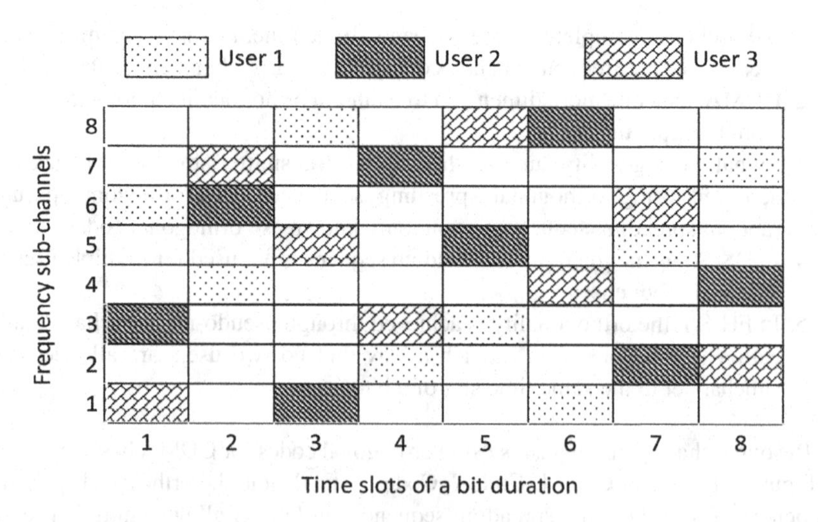

FIGURE 5.13 An example of subchannel assignment strategy for an FH-SS based CDMA system considering 8 subchannels and 8 time slots per bit duration and 3 users.

Example of CDMA through DS-SS

Let us see an example of CDMA through DS-SS. Consider 4-bit spreading sequences for two users, $c_1[n]$ and $c_2[n]$, as

$$c_1[n] = 0\ 0\ 1\ 1$$

$$c_2[n] = 0 \quad 1 \quad 0 \quad 1$$

When the above spreading sequences are represented in NRZ-polar scheme of line encoding, we obtain orthogonal codes as

$$c_1[n] = -1 \quad -1 \quad 1 \quad 1$$

$$c_2[n] = -1 \quad 1 \quad -1 \quad 1$$

Now, consider bit '1' is to be transmitted from user 1. The transmitted signal using the code $c_1[n]$ can be represented as

$$s_1[n] = -1 \quad -1 \quad 1 \quad 1$$

and bit '0' represented as '–1' using polar signalling is to be transmitted from user 2. The transmitted signal using the code $c_2[n]$ can be represented as

$$s_2[n] = 1 \quad -1 \quad 1 \quad -1$$

Furthermore, signals s_1 and s_2 are transmitted simultaneously in wireless media. So, the received signal is given as

$$r[n] = s_1[n] + s_2[n] = 0 \quad -2 \quad 2 \quad 0$$

For signal detection using the code of user 1, $c_1[n]$, the receiver detects

$$\hat{s}_1 = s_1[n] * c_1[n] = 0 \quad 2 \quad 2 \quad 0$$

and signal detection using the code of user 2, $c_2[n]$, the receiver detects

$$\hat{s}_2 = s_2[n] * c_2[n] = 0 \quad -2 \quad -2 \quad 0$$

To have final detection as bit '0' or bit '1' for user 1 and user 2, we use the sign of addition of all entries of $\hat{s}_1$ $\hat{s}_1$, respectively. It should be noted that for demonstration purposes, we consider a noise-less scenario in this example. If noise is present, the addition of entries in $\hat{s}_1$ and $\hat{s}_2$ would differ from that of '4' and '–4', respectively. The detection would be erroneous if the sign of the addition changes due to noise.

The above example considers the perfect synchronization among the signals received from both users during detection. If the signals are not in perfect synchronization, there would be some interference among the transmitted signals from multiple users. This interference would lead to erroneous detection at the receiver. So, user synchronization in addition to noise plays an important role in CDMA system performance.

Furthermore, the source of asynchronism in CDMA systems may not be from the asynchronous clocks among users. Synchronization may also be lost in case the signals travel different distances while reaching the receiver from the transmitter. This is a common problem because the users are distributed in a geographic area, say, a cell in cellular communication systems. So, even if two users transmit their signals using a synchronous clock, the receiver receives signals from these users in asynchronous timing due to the difference of distances between the users and the receiver.

5.6.4 Noise Immunity of CDMA Systems

By now we know that CDMA and spread spectrum are complementary names of each other. CDMA usually refers to multiple access scheme using spread spectrum technology. The noise immunity, mainly for DS-SS based CDMA systems, can be explained keeping the following properties in view:

- Multiplying the message signal with high chip rate pseudo-random signal spreads the spectrum of the message signal.
- Multiplying the spread spectrum signal with high chip rate pseudo-random signal brings back the original spectrum of the message signal.
- When the spectrum is spread to a wider bandwidth, the power spectral density reduces to conserve the total power in the signal.

Similarly, a noise affected DS-SS signal when multiplied by the spreading sequence spreads the spectrum of the noise signal. Thus, the poser spectral density of the noise reduces while that of the signal enhances. So, this is a scenario in which low power spectral density noise affects a high power spectral density signal. So, the signal would not be experiencing much effect of noise. Hence, DS-SS systems are considered immune to noise and narrowband interferences.

5.7 SUMMARY

- In this chapter, we discussed communication systems in which either the message signal or the carrier signal is represented in digital form. This digital form may or may not be in binary representation.
- Baseband and passband signals and systems are differentiated based on the spectrum of the signal. Baseband signals are those with spectrum centered around or near zero frequency. Passband signals are those with spectrum centered away from zero frequency.
- All carrier modulated signals are passband signals. Carrier modulation allows multiplexing of signals from different users by modulating them with different

carrier frequency. This kind of multiplexing is known as 'frequency division multiplexing'.

- Baseband signals can be multiplexed only if they are in discrete time form using time division multiplexing.
- Some baseband modulation techniques for analog message signals are pulse amplitude modulation (PAM), pulse width modulation (PWM), and pulse position modulation (PPM). If the message signal is in digital form, we get corresponding digital baseband modulation.
- PAM is pulsed form derived from amplitude modulation, PPM is pulsed form derived from phase modulation, and PWM is pulsed form derived from frequency modulation.
- Modulation schemes which modulate message signals in digital representation are known as 'digital modulation schemes'.
- Digital modulation schemes corresponding to their analog modulation schemes are amplitude shift keying (ASK), phase shift keying (PSK), and frequency shift keying (FSK).
- All digitally modulated symbols can also be represented in vector form using orthogonal basis functions.
- Amplitude shift keying is a one-dimensional modulation scheme.
- Phase shift keying is a 2D modulation, and frequency shift keying is M dimensional modulation scheme where M is the order of modulation.
- PSK and FSK are constant amplitude modulation schemes and also referred to as 'constant power modulation schemes'.
- Using the vector representation, digitally modulated symbols can be represented graphically. This graphical representation of vector form of digitally modulated symbols is referred to as 'constellation'.
- Noise affects the constellation diagram by shifting the constellation points by the amount of noise added. The stronger the noise, the higher the displacement of a constellation point from its original location.
- Spreading the spectrum of a message signal before/after modulation is also one of the techniques to transmit signals. This technique is known as 'spread spectrum technique'.
- High chip rate spreading signal is used to spread the spectrum of the message signal.
- Spread spectrum technique is of two types: direct sequence spread spectrum and frequency hopping spread spectrum.
- Using spread spectrum technique, multiple users can be served at the same frequency simultaneously using orthogonal spreading codes. This technique is referred to as 'code division multiple access' (CDMA).
- DS-SS is inherently immune to noise and narrowband interference.

References

1. F. A. Aliev and L. Ozbek. Evaluation of Convergence Rate in the Central Limit Theorem for the Kalman Filter. *IEEE Transactions on Automatic Control*, 44(10), 1905–1909, 1999.

2. D. Dehay, J. Leskow, and A. Napolitano. Central Limit Theorem in the Functional Approach. *IEEE Transactions on Signal Processing*, 61(16), 4025–4037, 2013.
3. H. Fischer. *A History of the Central Limit Theorem: From Classical to Modern Probability Theory. Sources and Studies in the History of Mathematics and Physical Sciences*, Springer, New York, 2010.
4. J. Hu, W. Li, and W. Zhou. Central Limit Theorem for Mutual Information of Large Mimo Systems with Elliptically Correlated Channels. *IEEE Transactions on Information Theory*, 65(11), 7168–7180, 2019.
5. J.G. Proakis. *Digital Communications*. Electrical engineering series, McGraw-Hill: Singapore, 2001.
6. Wim C. van Etten. *Introduction to Random Signals and Noise: Appendix A: Representation of Signals in a Signal Space*, pp. 215–227, John Wiley & Sons, Ltd.: Chichester, England, 2005.

6 Electromagnetism and Spectroscopy

6.1 INTRODUCTION

This chapter focuses on the use of electromagnetic radiation for various spectroscopy techniques and their applications in physics, chemistry, materials science, biology, and medicine.

6.2 SPECTROSCOPY TECHNIQUES

6.2.1 Absorption Spectroscopy

Radiation in different regions of the electromagnetic spectrum can be absorbed by a sample to yield the absorption spectrum, which is a plot of the intensity of absorption as a function of wavelength or frequency. This happens when the incident photon energy impinging on the sample equals the separation between the energy levels of the atoms or molecules making up the sample. These energy levels are quantized, and thereby the absorption spectrum of a given sample is characteristic of that unique molecular structure. This fundamental absorption process is the basis for all kinds of spectroscopy that we will discuss later, namely UV-VIS, infrared, and nuclear magnetic resonance (NMR) spectroscopy.

In absorbance measurements by instrumentation, the wavenumber of light entering the sample is the same as the one being detected after it leaves the material, although its intensity is attenuated. The Beer-Lambert law indicates that when the incident radiation traverses a given solution, its intensity undergoes a change, which is a measure of the concentration of the sample and is also related to the structure of the absorbing molecular species and the path length of traversal of the beam [1]. Consider a light beam of intensity I_0 hitting a sample and emerging with attenuated intensity I, then the absorbance A is defined by:

$$\mathrm{A} = \log\left(I_0/I\right) = \varepsilon\,\mathrm{C}\,\mathrm{l} \tag{6.1}$$

where log represents the base-10 logarithm, and I_0, I, ε and A are all functions of the wavenumber of the radiation involved. Here ε goes by the name of molar extinction coefficient and has dimensions of [L/mol·cm], while C gives the sample concentration in units of [mol/L] and l defines the length of path traversed by the light beam within the sample in units of [cm].

For analytical measurements, absorbance is obtained by comparing the intensity of light that passes through the sample being studied with a so-called 'reference' or 'blank'. The two measurements can be made one after the other in a single-beam

DOI: 10.1201/9781003213468-6

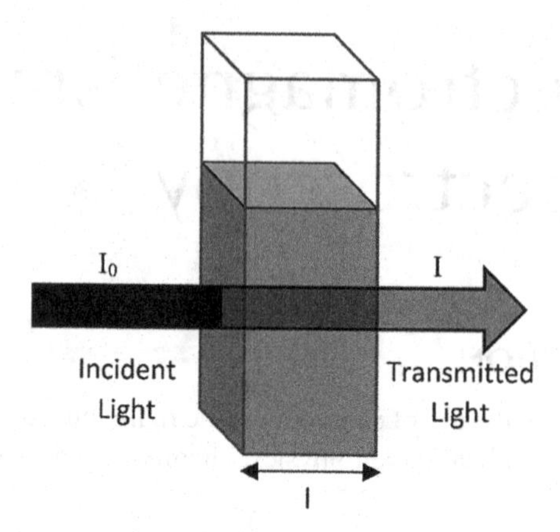

FIGURE 6.1 Diagram illustrating Beer-Lambert law.

instrument or simultaneously in a double-beam system. Later we will discuss how these pairs of measurements are made and the details of UV-VIS spectrophotometers, which are the preferred instrumentation to measure the absorbance of an unknown sample. The probability that a photon of given energy will be absorbed by a sample is enhanced by choosing a longer path length, higher concentrations and trying to match the energy of the incident radiation with the electronic transition occurring in the molecular sample. All these parameters are linked by the Beer-Lambert law cited above. The logarithmic nature of the Beer-Lambert law governing the ratio of light intensities (I_0/I) and the linear dependence of absorbance A on the sample concentration C governs the design and utilization of the UV-VIS instrumentation for analyte measurements [1,2].

Absorption spectroscopy is a powerful tool in analytical chemistry, atomic and molecular physics, and astronomy. We will discuss specific case studies in the UV-VIS section.

6.2.2 Emission Spectroscopy

6.2.2.1 Overview

Atoms and molecules can emit photons following excitation to a higher energy state by absorbing photons [3]. Such emission is referred to as 'luminescence' and categorized as either phosphorescence or fluorescence, which is determined by the excited state being a triplet state or singlet state, respectively. Excitation of atomic and molecular samples can also be accomplished by raising the temperature. One records the frequency or wavelength of photon(s) produced by the atomic or molecular species during their transfer from a higher energy state to a state of lesser energy, as illustrated in Figure 6.2. Fluorescence spectroscopy is complementary to absorption spectroscopy, in that fluorescence involves changeovers from a higher to lower energy state, whereas absorption deals with reverse transitions from lower to higher energy states [3,4]. Chemiluminescence results because of a chemical reaction.

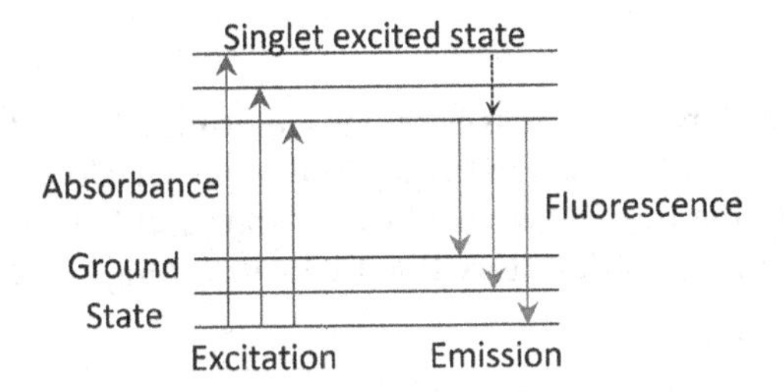

FIGURE 6.2 Diagram showing fluorescence.

Fluorescence spectroscopy is a more sensitive technique with lower detection limits than absorption spectroscopy and the photons emitted by fluorescent molecules have a lower wavenumber (energy) than the absorbed photons. The probability that a specific molecule in each excited state de-excites to the ground state is given by the parameter designated as quantum yield (Φ_f) [1]:

$$\Phi_f = \frac{k_f}{\left(k_f + k_{nf}\right)} \tag{6.2}$$

where k_f is known as the fluorescence rate constant and k_{nf} is termed as the rate constant for all non-fluorescent processes (e.g., phosphorescence, intersystem singlet to triplet crossing, and internal conversion from a singlet quantum state of greater energy to one of lower energy). If $\Phi_f = 1.00$, then every molecule that absorbs a photon and rises to an excited state fluoresces and emits a photon. To detect fluorescent molecules well, the higher the quantum yield the better the chances of detection. Like Φ_f, one can define a quantum yield (Φ_p) for phosphorescence.

6.2.2.2 Case Study

We will consider reactions between ions and molecules that occur due to collisions between species in both ground and excited states. Such processes have important ramifications for discharge plasmas, generation of soot in oxy-hydrocarbon flames, and the chemistry occurring in interstellar space [5]. For example, the presence of CHO^+ and $C_3H_3^+$ ions drives the chemi-ionization processes in oxy-hydrocarbon flames and is primarily responsible for the formation of abundant soot in such fuel-rich systems.

Ion-molecule collision reactions have been conducted in the laboratory in the energy range (0.1–0.9 keV) [5] and electronic emissions involving the radicals (CH and OH), ions (CH^+, CO^+ and N_2^+), and atomic hydrogen have been observed when positive ions (CHO^+ and H^+) undergo collisions with neutral molecules (CH_4 and N_2). A variety of emission cross-sections can be determined by these collision experiments, such as those involving excited CH species, N_2^+ ions and H_β transitions associated with H atoms.

In a typical ion-molecule collision experiment, there are projectiles (e.g., CHO^+ and H^+) that are produced by colliding electrons with methanol vapor in a flow

TABLE 6.1
A Summary of Emission Cross-Sections Determined for Atomic Hydrogen H_β, CH Radicals in the Excited State: CH(A) = CH($A^2\Delta$-$X^2\Pi$), Observed During the Collisions of CHO^+ Ions and CH_4 Molecules (P = 4 mTorr) in the Laboratory Kinetic Energy Range 100–900 eV

Kinetic Energy (eV)	Cross-section (10^{-20} cm^2)	
	H_β	CH(A)
900	0.2	1.02
750	0.8	3.70
650	1.24	6.32
550	0.76	5.62
450	0.12	8.70
300	–	4.40
100	-	2.74

Source: Reprinted from [5],

TABLE 6.2
A Summary of Emission Cross-Sections Determined for Atomic Hydrogen H_β, & CH Radicals in the Excited States: CH(A) = CH($A^2\Delta$-$X^2\Pi$) and CH(B) = CH($B^2\Sigma^-$-$X^2\Pi$), Observed During the Collisions of H^+ Ions and CH_4 Molecules (P = 4 mTorr) in the Laboratory Kinetic Energy Range 100–900 eV

Kinetic Energy (eV)	Cross-section (10^{-20} cm^2)		
	H_β	CH(A)	CH(B)
900	2.0	1.6	0.35
750	4.6	4.4	0.38
650	5.8	6.5	0.95
550	7.2	6.8	0.90
450	8.2	8.3	1.87
300	7.5	8.7	0.70
100	4.2	2.0	0.40

Source: Reprinted from [5],

chamber and removed via a small orifice in the anode [5]. An accelerating voltage propels the ions through a magnetic mass selector into a secondary chamber containing the neutral target molecules (CH_4 and N_2). Subsequently, a Faraday cup collects the emanating ions to facilitate measurement by a delicate electrometer and a gauge monitors the pressure of the target gas. Photons resulting from the ion-molecule collisions are viewed orthogonally to the ionic beam and wavelength-resolved using a scanning monochromator and then collected by a photomultiplier tube (PMT). The amplified PMT signal is relayed to a multichannel analyzer (MCA) for data storage, interpretation, analysis, and display.

Let us represent the CH radical in the ground state by the symbol CH(X). We can then show the formation of ionic CHO^+ by the so-called associative ionization reaction as follows: $CH(X) + O \rightarrow CHO^+ + e^-$, which occurs with a rate constant $k = 2.3 \times 10^{-13}$ $cm^3\ mol^{-1}\ s^{-1}$ over a temperature range (2000–2400 K) [5].

The emission cross-sections determined from such ion-molecule reactions can lead to better understanding of soot formation in oxy-hydrocarbon flames and lead to improved models for gaining enhanced insight into interstellar chemistry as well.

6.2.3 Ultraviolet-Visible Spectroscopy

6.2.3.1 Overview

Ultraviolet-visible (UV-VIS) is a particular kind of spectroscopy driven by electronic transitions involving atoms and molecules where the transitions lie in the ultraviolet and visible portions of the electromagnetic spectrum. In particular, the electrons that are tied to molecules with bonding and non-bonding orbitals absorb the UV-VIS energy and rise to higher energy antibonding molecular orbitals. As a result, there is a connection between the wavelengths of the absorption peaks and the kinds of bonds in a molecule, which aids in the identification of the active functional groups in the molecule that can be targeted for specific applications.

The Beer-Lambert law governs the absorbance of a sample as a function of concentration, calibration curve, and molar extinction coefficient.

Figure 6.3 illustrates the optical layout of a common UV-VIS spectrometer. The UV-VIS technique is used in a variety of analytical areas, which include the chemistry associated with transition metal ions, organic compounds, and biologically relevant macromolecules. In addition, the UV-VIS approach also finds applications related to the semiconductor industry, which among others focuses on the determination of the optical characteristics and thickness of thin films on specific substrates, together with the concomitant extinction coefficients and refractive indices of materials as a function of wavelength that provide insight into the chemical kinetics of relevant reactions.

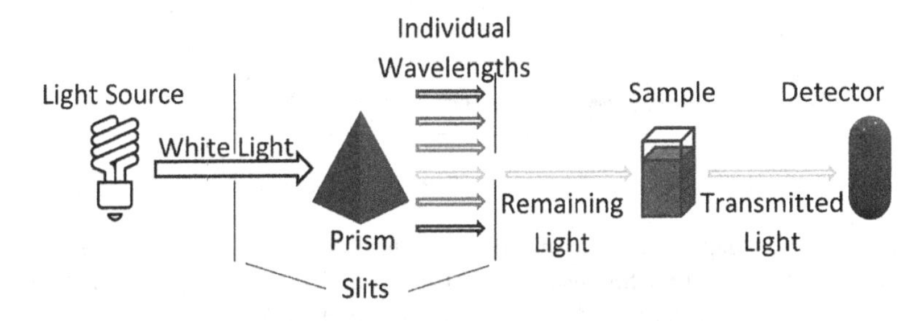

FIGURE 6.3 Experimental set-up of a typical UV-VIS spectrometer.

6.2.3.2 Case Study

UV-VIS data on thin films of tin dioxide (SnO_2) allow the calculation of the bandgap energy. The photon energy, E, can be calculated using the equation:

$$E = h\nu \tag{6.3}$$

where h is the fundamental Planck's constant, and ν gives the frequency of the radiation. Another quantity, α, can be obtained as the ratio of the absorbance and the thickness of the film. Finally, if $E^2\alpha^2$ is plotted against E, the result terminates in a line. When the line graph is extrapolated to intersect the X-axis, the X-intercept determines the energy of the bandgap for the thin film sample of SnO_2 [6].

As shown in Figure 6.4 and Table 6.3, the bandgap of SnO_2 film of thickness 78 nm has bandgap energy of around 3.458 eV. Then it decreases with thickness of 96.5 nm to around 3.383 eV. The higher thicknesses of the 373 and 908 nm films show an increase in the bandgap energy to 3.562 eV and 3.519 eV, respectively [6].

As compared to the SnO_2 bulk form (i.e., 3.6 eV), the bandgap energy for the film form in general is shifted to lower energy, which may have been caused by oxygen vacancies due to defect states present in the bandgap. The oxygen vacancies in the bandgap occupied higher states that reached the Fermi level. These oxygen vacancies affect the electronic density of state in the bandgap region. Therefore, the oxygen vacancy densities of the SnO_2 film are greater than that in the SnO_2 bulk form.

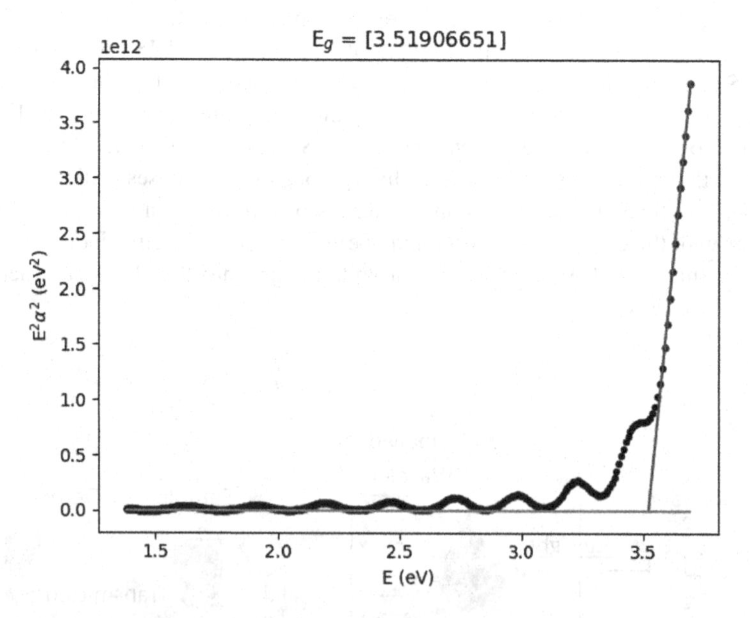

FIGURE 6.4 Bandgap energy determination of a thin film of SnO_2 on quartz. (Reprinted/adapted by permission from Springer Nature Customer Service Centre GmbH: Springer, *Contemporary Nanomaterials in Material Engineering Applications* by Mubarak N.M., Khalid M., Walvekar R., Numan A. (eds.), copyright (2021).)

TABLE 6.3
Summary of Bandgap Energies for a Variety of Thin Films of SnO_2 on Quartz

Thickness (nm)	Bandgap Energy (eV)
78	3.458
96	3.383
373	3.562
908	3.519

Source: Reprinted/adapted by permission from Springer Nature Customer Service Centre GmbH: Springer, *Contemporary Nanomaterials in Material Engineering Applications* by Mubarak N.M., Khalid M., Walvekar R., Numan A. (eds), copyright (2021).

6.2.4 Fourier Transform Infrared Spectroscopy

6.2.4.1 Overview

The technique of Fourier transform infrared (FT-IR) spectroscopy can be effectively deployed to obtain pertinent information relating to the molecular structure of both inorganic and organic materials. Typically, infrared radiation in mid-IR range (400–4000 cm^{-1}) is absorbed by a sample material and the FT-IR spectrum obtained in terms of intensity of absorbance as a function of the wavenumber (cm^{-1}).

The FT-IR spectra of the samples were collected using the PerkinElmer Frontier FT-IR/NIR spectrometer shown in Figure 6.5 and utilizes the so-called universal attenuated total reflection (UATR) polarization accessory [6]. The UATR device incorporates a DiComp™ crystal, which in turn includes a diamond ATR and a focusing component made up of zinc selenide (ZnSe).

The advantage of using the ATR accessory is that the sample preparation time is minimized by utilizing the diamond crystal and other components (see Figure 6.6 (a–c)). The way the ATR accessory works is that it accumulates the internally reflected mid-IR radiation coming from the optically dense diamond crystal, which is in direct contact with the material sample. Therefore, the energy carried by the

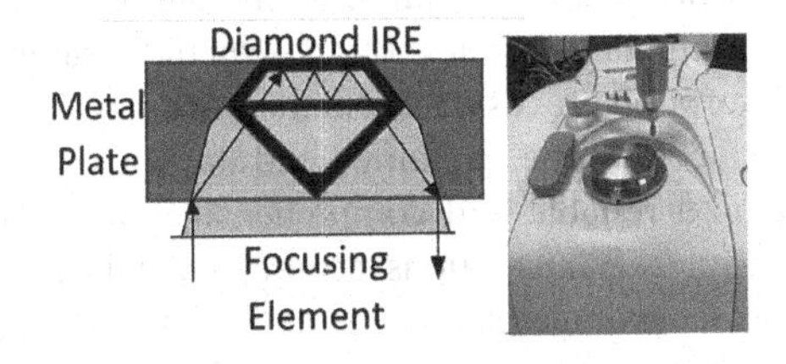

FIGURE 6.5 (left) Schematic of a UATR top plate and (right) PerkinElmer Frontier FT-IR/NIR spectrometer with the UATR accessory. (Reprinted from [7]. Copyright (2021), with permission from Elsevier.)

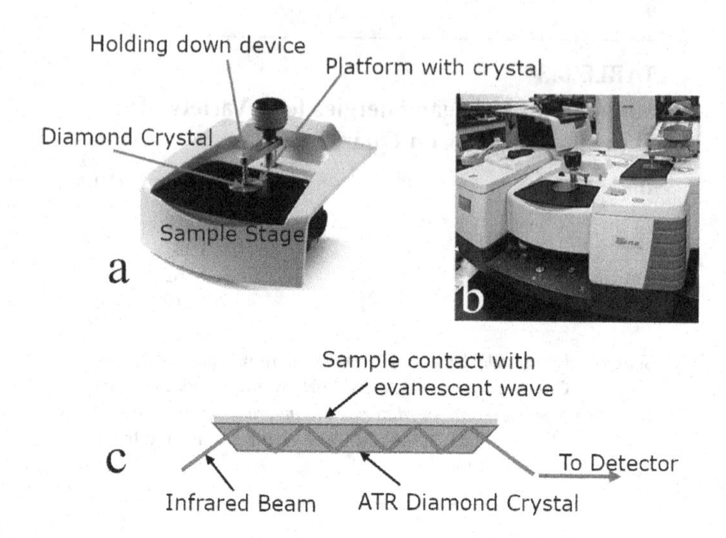

FIGURE 6.6 (a) ATR sample stage, (b) Thermo Scientific™ Nicolet™ iS50 FT-IR spectrometer, and (c) diagram showing the inner workings of the ATR top plate. (Reprinted/adapted by permission from Springer Nature Customer Service Centre GmbH: Springer, *Contemporary Nanomaterials in Material Engineering Applications* by Mubarak N.M., Khalid M., Walvekar R., Numan A. (eds.), copyright (2021).)

evanescent wave (see Figure 6.6 (c)) is absorbed by the sample, and the energy of the wave is diminished. The corresponding attenuated energy is measured by a detector, and the instrument then records the FT-IR absorption spectrum of the sample.

FT-IR spectroscopy has been used effectively to monitor measurable small variations in absorbance corresponding to sub-parts per million (sub-ppm) levels of lunar analog mineral samples [8] and atmospheric molecular species (e.g., HCl, SO_2, NO_2 and NH_3) adsorbed on a variety of tubing made of different materials (e.g., aluminum, copper, stainless steel, and Teflon™) and under a wide range of pressure and temperature conditions.

The absorption spectra were measured in the mid-IR range (400–4000 cm^{-1}) using a Nicolet Magna-550 FT-IR spectrometer, in tandem with a 10-m multipass absorption cell, plus a deuterated triglycine sulfate (DTGS) detector [9]. The assortment of tubing materials tested were motivated by the choice of materials used in constructing hardware used for handling and measuring samples of atmospheric trace gases that are important for a better understanding of climate modeling, simulations, and chemical kinetics involving these species. By making the assumption of a monolayer associated with the adsorption process, the amount of adsorption for combinations of various gas-solid interfaces can be quantified and the corresponding rate constants determined. To develop a reliable and comprehensive adsorption isotherm model incorporating the gas-surface interactions accurately, it is important to precisely determine the adsorption coverage parameters and the so-called Langmuir rate constants for a variety of tubing material-trace gas combinations by measuring residence times and obtaining the associated activation energies.

The absorption features of the trace gas constituents were investigated with the FT-IR spectrometer by recording the individual rotational-vibrational (ro-vibrational) spectral features of the molecule, thereby permitting the correct assignment of transitions to be made for the different trace species examined at various concentration levels (5–100 pm). The FT-IR spectroscopic technique facilitated the indirect observation of the adsorption effects due to surface interactions of very low concentrations of highly polar molecules, such as HCl, with the interacting surfaces of the measuring apparatus. The adsorption 'sticking' effects of the HCl onto the plumbing and the inner walls of the measuring apparatus provided impetus for this work. The development of accurate reference spectral databases from field measurements using the FT-IR spectrometer usually requires a continuous gas flow, such as the approach used by the National Institute of Standards and Technology (NIST) organization in the United States. The constant flow of known concentrations of primary standard gas mixtures through sample cells minimizes biases in the determination of the concentration caused by adsorption of the sample onto the cell walls, which in turn minimizes the reported sample concentration loss due to adsorption. A systematic study of the interaction of atmospheric trace gases, such as HCl, NO_2, NH_3, and SO_2 with sheets of various tubing materials that are used for the development of gas handling systems, can aid designs that minimize the effects of adsorption. Atmospheric monitoring of trace gases can lead to improvements on the techniques of correcting the measured concentration by means of fitting Langmuir adsorption isotherm curves to the observed data. Tubing materials of interest, such as stainless steel, Teflon, aluminum, and copper were used to investigate the effects of adsorption [9].

6.2.4.2 Experimental

Accurate measurements of the absorption spectra were performed using the FT-IR spectrometer and associated accessories. Figure 6.7 shows the experimental layout of the Nicolet Magna-IR 550 spectrometer [9]. The resolution of the FT-IR instrument was varied from 0.25 to 32 cm^{-1} as deemed appropriate for the measurements. In

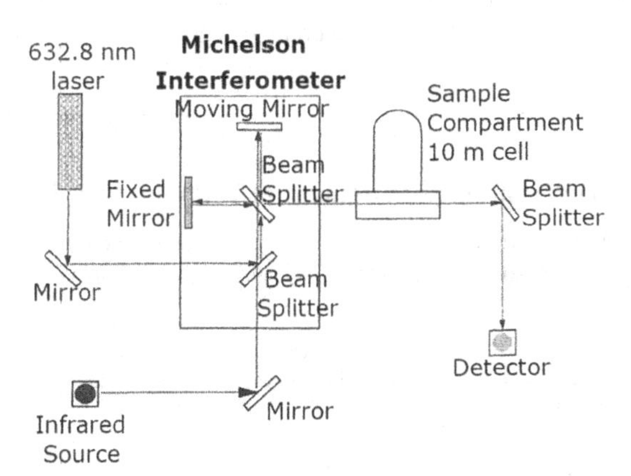

FIGURE 6.7 The Nicolet Magna-550 FT-IR spectrometer experimental arrangement [9].

addition to a DTGS detector at room temperature, a mercury cadmium telluride (MCT) liquid nitrogen cooled detector was used to record absorption spectra in the range 10–500 ppm, requiring fewer than 64 scans (10 minutes). A sensitive MCT-A detector and KBr windows were used throughout for high transparency in the mid-IR region.

6.2.4.3 Case Study

For illustrative purposes, we choose the interaction of gaseous NH_3 on aluminum. Aluminum sheets with total surface area of 1300 cm^2 were exposed to 760 Torr of the gas concentration of NH_3 in N_2. The absorbance as a function of time was recorded using the FT-IR instrument by taking snapshots of the absorbance at the fundamental line $v_2(a_1) = 967.31$ cm^{-1}. The results for the spectroscopic data scan for the NH_3-aluminum adsorption reaction are given in Table 6.4 and plotted in Figure 6.8 [9]. The time varying concentration corresponding to the absorbance was calculated from the Beer-Lambert expression

$$\ln\left(\frac{I}{I_0}\right) = -\sigma c N_0 l \tag{6.4}$$

and the absorption cross-section σ determined.

6.2.5 Raman Spectroscopy

Raman spectroscopy is a powerful non-invasive technique facilitated by a laser light source interacting with a material sample. Most of the light scatters off the sample at the same wavenumber as the incident radiation and is termed 'Rayleigh scattering'. In addition, a small but detectable amount of incident radiation (appropriately filtered to remove the Rayleigh portion) experiences a wavenumber shift and results in what is known as 'Raman scattering'.

The sample vibrates uniquely to its structure, and each vibration mode uniquely alters the emitted photon wavelength; that change is graphed as intensity versus wavenumber to yield a Raman spectrum. Important applications of Raman spectroscopy

TABLE 6.4
NH_3-Aluminum Desiccator Measurement at 100 ppm NH_3 in N_2 [9]

Ps(torr)=760	Initial Concentration		Reaction Volume	V(cm^3)=7500
Pa(torr)=30	Res(cm^{-1})=0.5	c(ppm)=100 T(°C)=5	Surface Area	A(cm^2)=1300
Time (min)	**Time (sec)**	**Absorbance at $v_2(a_1)$= 967.31 cm^{-1}**	**Number Absorbed/cm^2 (calculated)**	**Instantaneous Concentration (ppm)**
2.0	120	0.0187	0.00	100.0
16.0	960	0.0110	5.89(14)e+15	58.8
29.0	1740	0.0091	7.34(17)e+15	48.7
43.0	2580	0.0081	8.11(19)e+15	43.3
125.0	7500	0.0051	1.04(2)e+16	27.3
204.0	12240	0.0042	1.11(3)e+16	22.5

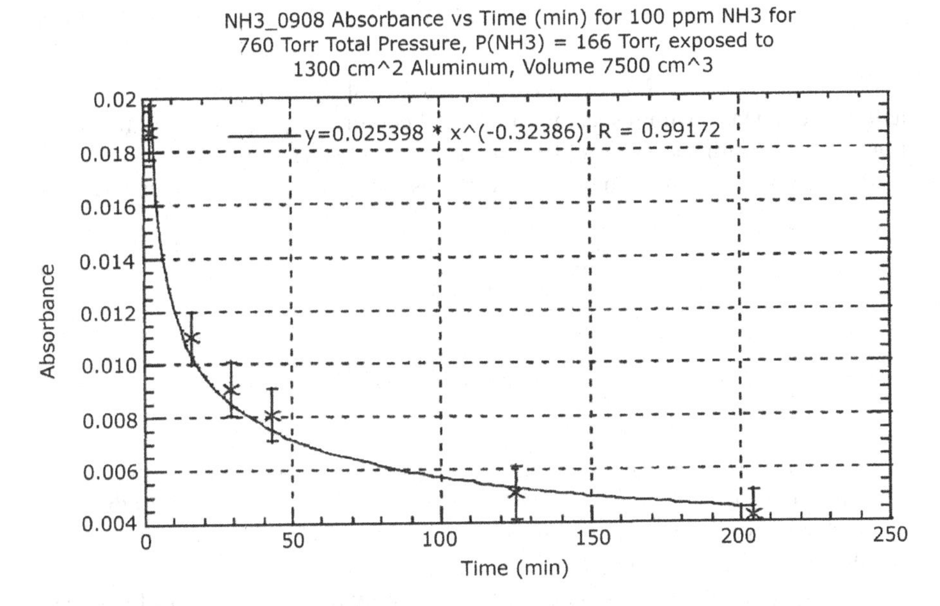

FIGURE 6.8 Instantaneous absorbance for desiccator samples at 100 ppm for NH_3-Al interface reaction [9].

are in materials science, nanotechnology, detection of explosives at stand-off distances, solid state chemistry, biology, medicine, and preservation of art and cultural heritage.

Inelastic scattering mechanism and polarization effects are involved in the Raman shift. Besides using the Raman effect to characterize the vibrational features of molecules, rotation and other low-frequency modes can also be identified.

The unusual scattering phenomena exhibited by molecules, where the light bouncing off the sample has the same wavenumber, and somewhat shifted wavenumbers, was predicted in 1923 by the Austrian theoretical physicist Adolf Smekal (1895–1959). The effect was later confirmed experimentally by the Indian spectroscopist Sir C.V. Raman, in 1928, who was rewarded with the Nobel Prize in Physics in 1930, and to this day the scattering phenomenon to lower and higher wavenumbers (Raman shift) bears his name and is known as the 'Raman Effect'.

We provide here the classical theory of the Raman Effect where we treat the molecules in the form of an electric dipole of moment μ undergoing simple harmonic motion.

The incident light is considered in the form of a plane electromagnetic wave:

$$E = E_0 \cos\left(2\pi \nu_0 t\right) \tag{6.5}$$

where E_0, ν_0, and t are the amplitude, incident frequency, and time parameters, respectively. The induced electric dipole moment μ is directly proportional to the electric field E, where the proportionality constant is referred to as the polarizability coefficient α and given by:

$$\mu = \alpha E \tag{6.6}$$

The intensity of the Raman signal is a function of the derivative of the polarizability coefficient α with respect to the normal coordinate q representing the vibrations, which in turn has a sinusoidal form: $q = q_0 \cos(2\pi\nu t)$.

In the harmonic regime of molecular vibrations, the polarizability can be expanded as

$$\alpha = \alpha_0 + \left(\frac{\partial \alpha}{\partial q}\right)_0 q_0 \tag{6.7}$$

$$\mu = \alpha_0 E_0 \cos(2\pi\nu_0 t) + \frac{1}{2}\left(\frac{\partial \alpha}{\partial q}\right) E_0 \left[\cos\{2\pi(\nu_0 + \nu_m)t\} + \cos\{2\pi(\nu_0 - \nu_m)t\}\right] \tag{6.8}$$

In Equation (6.8), the first term on the right-hand side is representative of the Rayleigh signal from the induced dipole moment of the same frequency as the incident radiation, while the second term describes the anti-Stokes Raman shift $(\nu_0 + \nu_m)$ and the Stokes Raman shift $(\nu_0 - \nu_m)$, respectively.

Next, we present the quantum mechanical description of the Raman Effect. One simplifying assumption that is useful is to consider the light incident on the sample classically and treat it as a perturbation for the molecular sample undergoing Raman scattering.

If we consider P to be the induced dipole moment, and a transition from an initial energy eigenstate n to a final eigenstate m, then the induced transition matrix element P_{mn} is given by Equation (6.9). The intensity of the Raman signal is then determined by the square of P_{mn}.

$$P_{mn} = \int \psi_m^* P \psi_n d\tau \tag{6.9}$$

Here ψ_n and ψ_m are the initial and final wavefunctions of the molecular system, respectively, acting on the induced dipole moment P, and the integration is performed over all space.

By invoking the time-dependent Schrödinger equation, one obtains the transition moment matrix given by:

$$P_{mn} = \frac{1}{h}\sum_r \left(\frac{M_{nrM}{}^{rm}}{\nu_{rm-\nu^0}} + \frac{M_{nr}M_{rm}}{\nu_{rm} + \nu_0}\right) E \tag{6.10}$$

In Equation (6.10), h is Planck's constant, E is the electric field of the incident light, and M_{nr} is the transition moment connecting the induced virtual states and the unperturbed states (r) of the molecule under study. The square of the transition moment P_{mn} yields the intensity of the Raman signal w_{mn} given by Equation (6.11), where c is the speed of light in vacuum. Here $P_{nm}^2 = P_{mn}^2$.

$$w_{mn} = \frac{64\pi^2}{3c^2}(\nu_0 + \nu_{nm})^4 P_{mn}^2 \tag{6.11}$$

Figure 6.9 is a schematic illustrating the Rayleigh, Stokes, and anti-Stokes transitions from the quantum mechanics perspective, for the case electronic states and the induced virtual states are not in close proximity on the energy scale.

A Thermo-Scientific DXR SmartRaman Spectrometer with 780 nm laser excitation (with power setting of around 6 mW) and good beam quality ($M^2 \leq 1.5$) has been used to carry out the Raman spectroscopy investigations of these samples.

The experimental layout for the Raman experiments is shown in Figure 6.10. The schematic shows the laser beam being focused by lenses on the material sample housed in a temperature controlled Ventacon Cell that can be raised from room temperature up to 200°C. A notch filter in the path of the scattered light beam helps remove the Rayleigh component and passes through the Raman scattered radiation on to an imaging spectrometer and a CCD detector to record the Raman spectrum. For a slit-width of 25 μm, the resolution of the spectrometer was around 1.93 cm^{-1}. Depending on the nature of the sample material, the Ventacon Heated Cell facilitates loading of solid samples (e.g., on plates) on its surface or inside the interior volume

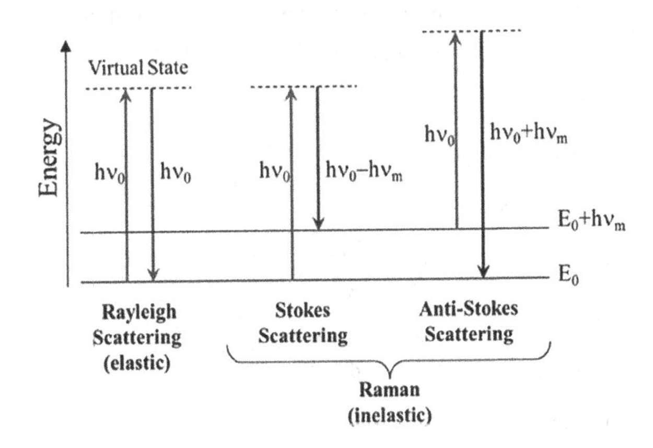

FIGURE 6.9 Quantum mechanical Raman spectroscopy schematic showing Stokes and anti-Stokes transitions [10].

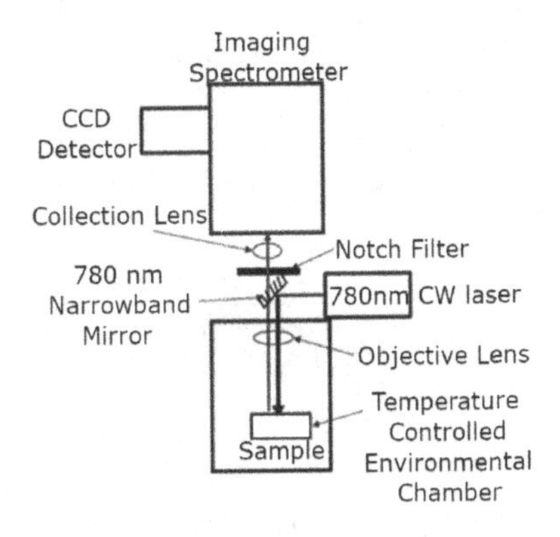

FIGURE 6.10 Experimental set-up for Raman spectroscopy [10].

(e.g., nanopowders and carbon nanotubes) and subsequently studying them at elevated temperatures.

6.2.6 X-Ray Spectroscopy

6.2.6.1 Overview

X-ray diffraction (XRD) is a non-destructive and informative method to characterize the crystal structure, crystallite size, and orientation of a material. The XRD measurements are based on Bragg's law, illustrated in Figure 6.11, and provide insight into the crystalline structure of the nanomaterial samples. The solid material samples studied were either a highly ordered array of atoms (crystalline) or had a randomly organized atomic pattern (amorphous). In the technique of X-ray emission spectroscopy, the X-rays excite electrons in the atomic shell of the sample, subsequently detecting the emitted photons. Both non-resonant and resonant categories of X-ray emission are possible.

The geometry of the experimental layout for X-ray diffraction studies, as illustrated in Figure 6.12, shows the relative orientations of the X-ray source, sample, and detector, where the extrapolation of the incident X-ray beam makes an angle θ with the sample and angle 2θ with the line-of-sight of the detector.

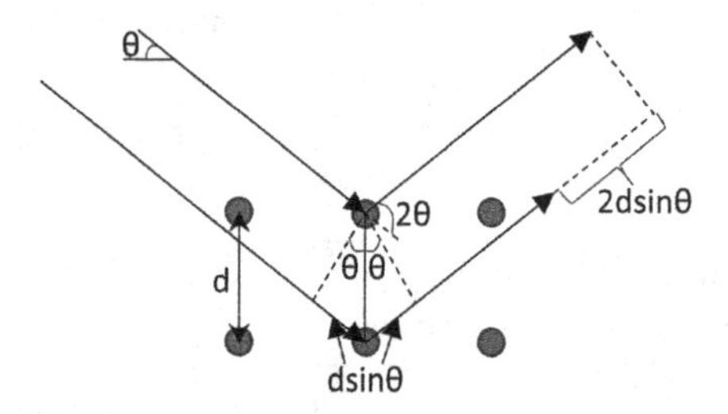

FIGURE 6.11 Illustration of Bragg's law.

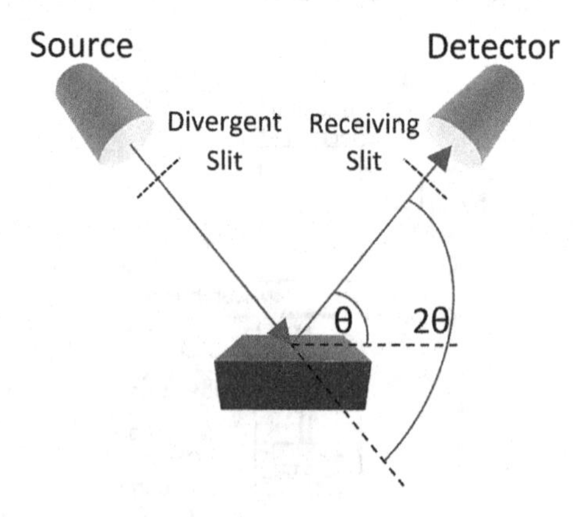

FIGURE 6.12 Schematic of X-ray diffraction [7,11].

6.2.6.2 Case Study

Next, we present a case study involving the recording of X-ray diffraction spectra of powder and film samples of tin dioxide (SnO_2). The instrument used for this purpose is a Thermo Fisher Scientific™ ARL™ EQUINOX 100 X-ray diffractometer (see Figure 6.13) [6,11]. The equipment incorporates a spinning stage, and the dimensions of the X-ray beam are approximately 5 mm × 300 μm. Owing to the special EQUINOX curved detector, one can record the diffraction peaks simultaneously over a wide range of 2θ values in real time, without any scanning requirements. Like any other XRD, the present instrument has an X-ray light source, which is accompanied by a sample stage, optical receiver, and detector. Both the X-ray tube and detector are stationary in the ARL EQUINOX instrument, while the sample stage is the only component that is moveable.

We illustrate in Figures 6.14 and 6.15 the XRD spectra of powdered SnO_2 collected over a range of 2θ values from 20 to 80°. We have been able to correlate the data with the corresponding Miller indices (*h k l*) at 26.34° (1 1 0), 33.60° (1 0 1), 38.25° (2 0 0), 51.56° (2 1 1), 55.01° (2 2 0), 62.13° (3 1 0), 64.97° (1 1 2), and 66.15° (3 0 1), respectively. Keep in mind that each set of indices is representative of the various faces of the crystalline shape of SnO_2. As a result of these measurements, we have been able to validate the rutile tetragonal structure (cf. Reference Card Number: JCPDS 41-1445).

Similarly, XRD data are collected from $2\theta = 0°$ to $2\theta = 80°$ for thin films of SnO_2 on UV-quartz substrate. The pure quartz substrate is compared to different thickness ratios of SnO_2 (i.e., 41, 78, 96.5, 373, and 908 nm). A broad peak is recorded for the bare quartz around 21.62°, where the broadening is between 10–35° and shows an amorphous characteristic. The lower thicknesses of SnO_2 (i.e., 41, 78, and 96.5 nm) show little evidence of the tetragonal rutile structure of SnO_2. As the thickness increases, the SnO_2 structural patterns start to appear. Film thickness of 373 nm

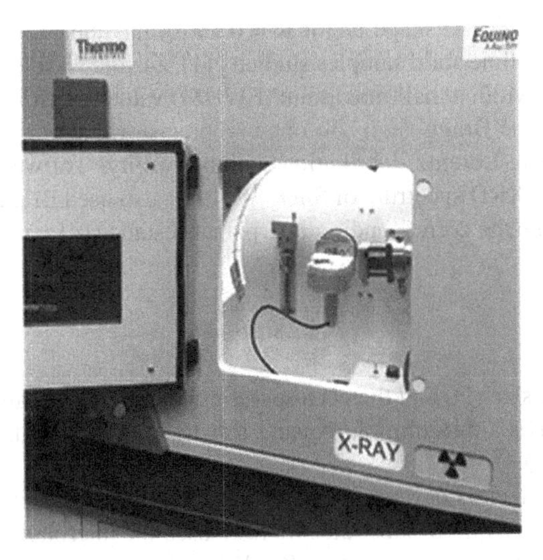

FIGURE 6.13 Thermo Fisher Scientific™ ARL™ EQUINOX 100 X-ray diffractometer [6].

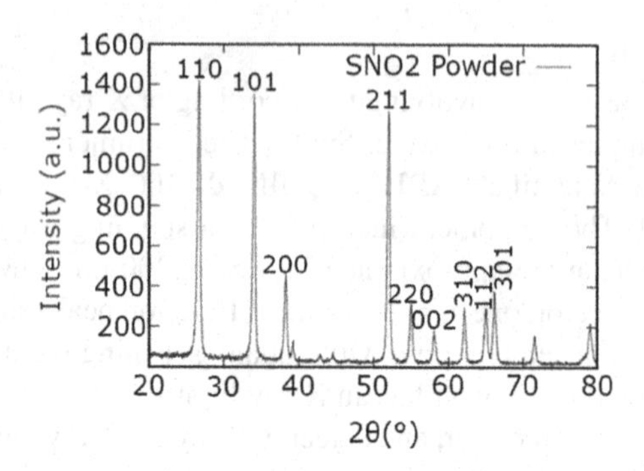

FIGURE 6.14 XRD spectra of SnO_2 in powder form. (Reprinted/adapted by permission from Springer Nature Customer Service Centre GmbH: Springer, *Contemporary Nanomaterials in Material Engineering Applications* by Mubarak N.M., Khalid M., Walvekar R., Numan A. (eds.), copyright (2021).)

records 26.36° (1 1 0), 33.92° (1 0 1), and 51.5° (2 1 1). Indeed, the highest film thickness of 908 nm shows more pronounced peaks at 26.33° (1 1 0), 33.63° (1 0 1), and 51.5° (2 1 1), which confirms the tetragonal rutile structure of the SnO_2 film on quartz. The film samples show the amorphous behavior due to the small size of particles compared to the powder SnO_2 form [11].

It is clear from the XRD data for an individual SnO_2 crystal that the (1 1 0) plane correlates with the face of least energy and largest surface area, whereas (1 0 1) and (1 0 0) planes correspond to smaller faces. The Debye-Scherer equation (6.12) enabled the determination of the crystallite size (D) of the SnO_2 samples.

$$D = \frac{K\lambda}{\beta \cos\theta} \tag{6.12}$$

In Equation (6.12), the shape factor K = 0.98 for the choice of cubic unit cell of the spherical crystalline solid samples studied [11], λ is the wavelength of the X-rays used, β is the full width at half maximum (FWHM) value determined from the spectral line, and θ is the Bragg angle. To obtain a precise value of the parameter β, we utilized the Cauchy-Lorentz distribution within the Fityk software to fit the sharp peak (1 1 0) in the XRD spectrum of SnO_2. We have also used Bragg's law (Equation (6.13) to determine the corresponding interplanar distance (d).

$$d = \frac{\lambda}{2\sin\theta} \tag{6.13}$$

The crystallite size (D) and the corresponding interplanar distance (d) of SnO_2 in powder form and as a thin film on a quartz substrate have been determined and are shown in Table 6.5.

The lattice parameters have been calculated from XRD parameters for SnO_2 powder and thin film on a quartz substrate. We have used the interplanar spacing formula for a tetragonal unit cell as in Equation (6.14).

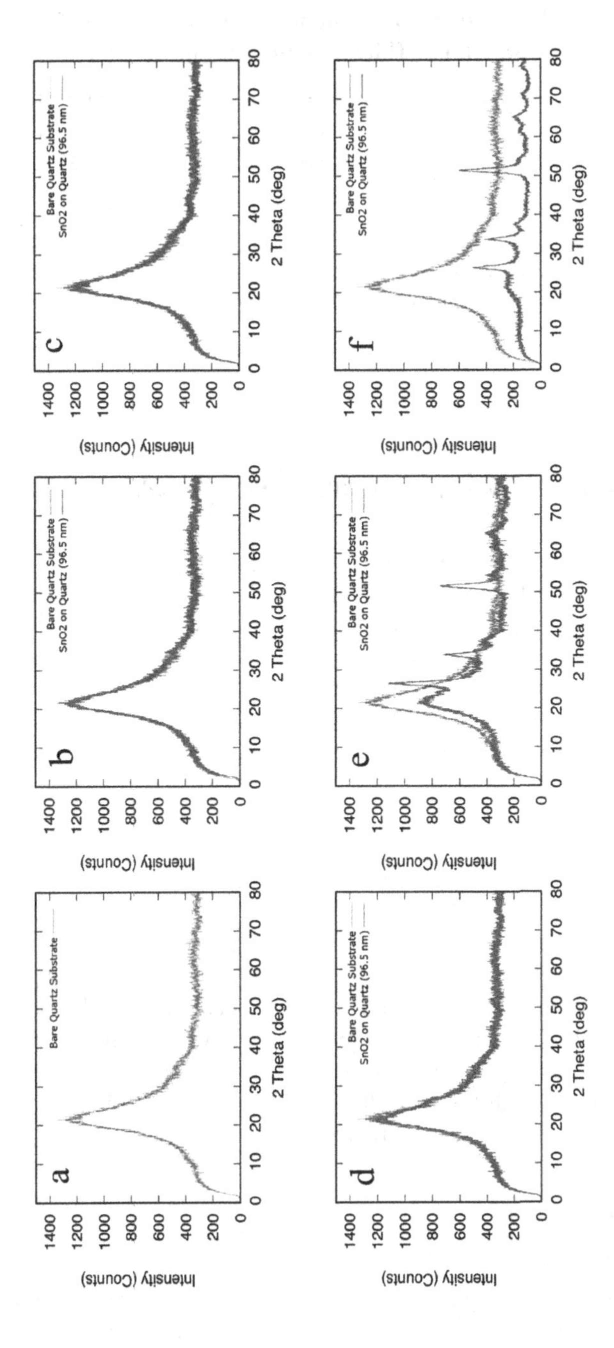

FIGURE 6.15 XRD spectra of different thicknesses of SnO_2 films on UV-quartz substrate. (Reprinted/adapted by permission from Springer Nature Customer Service Centre GmbH: Springer, *Contemporary Nanomaterials in Material Engineering Applications* by Mubarak N.M., Khalid M., Walvekar R., Numan A. (eds.), copyright (2021).)

TABLE 6.5
Calculated Crystallite Size (D) and Interplanar Distance (d) of SnO_2 in Powder form and as a Thin Film of Thickness 908 nm on Quartz

Sample	$2\theta°$	FWHM ($2\theta°$)	Crystallite Size (D) (nm)	Interplanar Distance (d) (Å)
SnO_2 powder	(1 1 0) 26.34	0.424	19.044	3.380
SnO_2/quartz (908 nm)	(1 1 0) 26.33	1.272	6.7	3.382

Source: Reprinted/adapted by permission from Springer Nature Customer Service Centre GmbH: Springer, *Contemporary Nanomaterials in Material Engineering Applications* by Mubarak N.M., Khalid M., Walvekar R., Numan A. (eds.), copyright (2021).

TABLE 6.6
Calculation of Lattice Parameters from XRD of SnO_2 Powder and Thin Film on Quartz

Sample	$2\theta°$	(*hkl*)	Interplanar Distance (d) (Å)	Lattice Parameter (a = b) Å	Lattice Parameter (c) Å
SnO_2 powder	38.25	(2 0 0)	2.351	4.702	
	58.1	(0 0 2)	1.586		3.173
SnO_2/quartz (908 nm)	37.57	(2 0 0)	2.392	4.784	
	58.31	(0 0 2)	1.581		3.1623

$$\frac{1}{d^2} = \frac{h^2 + k^2}{a^2} + \frac{l^2}{c^2} \tag{6.14}$$

where d is the interplanar distance as calculated in Table 6.5; h, k, l are the Miller indices, and a and c are lattice constants (i.e., $a = b$, c). To calculate the a parameter, the second part of the equation (6.14) must vanish. Therefore, we choose the peak that corresponds to (2 0 0) plane with $2\theta = 38.25°$ for powder form and $2\theta = 37.57°$ for the thin film. To calculate the lattice parameter c, we choose the (0 0 2) plane and its corresponding angles of diffractions $2\theta = 58.1°$ and $2\theta = 58.31°$ for the powder and film forms, respectively, as seen in Table 6.6.

6.2.7 Mass Spectrometry

6.2.7.1 Overview

Mass spectrometry is a technique that determines the mass-to-charge ratio of ions and displays as a mass spectrum in terms of a graph of the ion signal intensity versus its mass-to-charge ratio. The technique is used by scientists for obtaining the relative masses and isotopic abundance of ions [12].

Mass spectrometry is a method that is used to determine the composition of pure samples or mixtures. By separating the mixture into its different components through a variety of means (such as gas chromatography), it is possible to match the unknown composition to known compound(s) by using a spectral library (example of such application can be seen in Figure 6.16) [13–17].

A mass spectrometer is composed of an ion source which emits the ions in the presence of an external magnetic field, similar to the arrangement illustrated in Figure 6.17. The magnetic field separates the ions by mass-to-charge ratios.

In the experimental arrangement illustrated, the first step is to vaporize the atoms/molecules of the sample, followed by ionization that generates a positive ion by removing an electron. A stream of positive ions is accelerated by a potential difference, V, present in the region between the ion source and the positive metal plate (as shown in Figure 6.17). These positive ions emanate from an orifice in the plate to enter a region of constant magnetic field oriented in the orthogonal direction, traverse

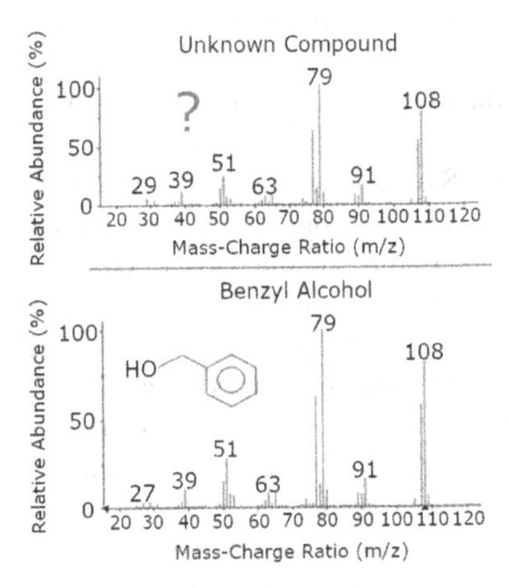

FIGURE 6.16 Comparison of mass spectrum of an unknown compound and benzyl alcohol [18].

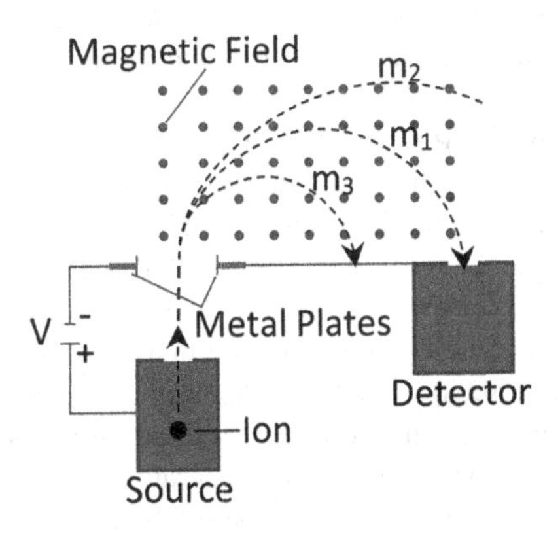

FIGURE 6.17 Schematic of a mass spectrometer.

semicircular paths, and impinge on the detector. Only specific positive ions with the correct radius for the circular path enter the detector and are counted per second.

The semicircular paths traced by the ions as they pass through the magnetic field allow us to relate the Lorentz force law $F = qvB \sin\theta$ to the centripetal force $F = \frac{mv^2}{r}$, whereby

$$qvB \sin\theta = \frac{mv^2}{r} \Rightarrow \frac{mv}{qr} = B, \tag{6.15}$$

with $v = \sqrt{\frac{2qV}{m}}$

$$\frac{m}{q} = \frac{B^2 r^2}{2V} \tag{6.16}$$

In this way, we can selectively choose which ions will reach the mass spectrometer [12–17].

In another variant of the mass spectrometer instrumentation, a time-of-flight measurement is used to obtain the charge-to-mass ratio of an ion [14,15]. In time of flight mass spectrometry (TOFMS), the idea is to use an electric field of given magnitude to accelerate ions in such a way that ions having the same charge will possess identical kinetic energies. As a consequence, the mass-to-charge ratio of the ions will dictate their velocities, whereby more massive ions with the same charge will acquire lower speeds and ions with higher charge will be able to travel at enhanced velocities. One is then able to determine the time it takes for a given ion to travel a known distance to the detector, which in turn is a function of its velocity and thereby its mass-to-charge ratio as well. A determination of the charge-to-mass ratio of the ion helps to identify the ion being studied.

Electric and magnetic fields are used to design traps to capture ions in an isolated system [16]. Such ion traps find applications in fundamental physics research, as well as in mass spectrometry and quantum information science. A pair of popular ion traps are the so-called 'Penning and Paul traps', where the former deploys a combination of electric and magnetic fields to create a potential well, and the latter utilizes a combination of static and oscillating electric fields to produce a well.

6.2.7.2 Gas Chromatography-Mass Spectrometry

The hyphenated gas chromatography-mass spectrometry (GC-MS) analytical technique combines the strengths of gas chromatography and mass spectrometry to identify components within a given sample [17]. As illustrated in Figure 6.18, a flowing carrier gas transports a mixture of vaporized molecules into a capillary column, separates them into components as a function of their mass, and passes them into the mass spectrometer to obtain a 3D spectrum that displays abundance versus the retention times of the components of the mixture and the charge-to-mass ratio of the ions. As a result, a cross-section of the 3D image can generate either a 2D chromatogram or a 2D mass spectrum, and thereby one can separate the components of a mixture and identify them using the combined GC-MS technique.

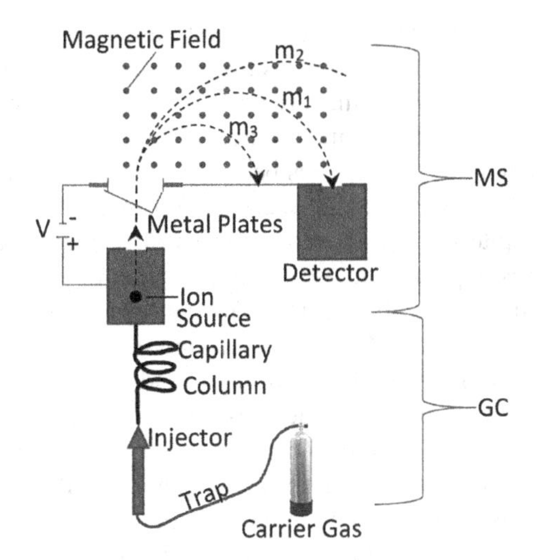

FIGURE 6.18 Schematic of GC-MS process.

The versatile GC-MS analytical method finds a myriad of applications, including detection of explosives and drugs, investigation and spreading of wildfires, analysis of analytes for environmental protection, and for the detection and identification of organic molecules on the red planet, Mars, as part of an ongoing NASA mission (e.g., the Mars Science Laboratory [MSL] aboard the Curiosity rover). Another area of widespread usage is in airport and shipyard security to detect hazardous chemicals and illicit drugs in luggage, on individuals, and in cargo. The GC-MS technique also finds application in detecting and identifying trace amounts of materials in artifacts thought to have been damaged or disintegrated and unsalvageable. Hyphenated techniques such as liquid chromatography-mass spectrometry (LC-MS) and GC-MS are highly sensitive analytical tools that allow for detection of even trace miniscule amounts of a material.

6.2.7.3 Case Study

The case study introduced for the application of the GC-MS technique is the contamination problem posed by the outgassing of materials used by spacecraft for planetary missions, especially relating to *in situ* analysis of organic molecules. A series of in-depth studies have emphasized the critical issue of outgassing of such polymeric materials, namely, e.g., [19–22]. In addition, pointers gathered from terrestrial geochemical investigations relating to contamination control, along with ocean floor and glacial ice core studies, help to plan and guide future space missions [23–25].

To deal with the critical issue of organic contamination, a NASA commission termed the Organic Contamination Science Steering Group (OCSSG), which was formed prior to the solicitation for the MSL mission aboard the Curiosity rover. Mahaffy et al. [26] is a good resource for the strategies crafted by the Steering Group for the NASA Engineering and Operations teams to keep in mind while designing, fabricating, assembling, and operating the Mars Lander systems, together with the science team overseeing the analysis of the critical data received from MSL. In fact, some of the same strategies have been effectively used during the Viking mission [27]. It must be kept in mind that all kinds of organic materials (e.g., gasses, particulates, films, etc.), especially those that are volatile, can contribute to the contamination problem by becoming a potential source of organics. For instance, this can happen when the contaminant molecules become either an integral part of the rover or the MSK aeroshell due to processing and environmental issues associated with the flight hardware. Despite mitigation efforts, traces of residual contamination often remain; therefore, it is imperative to use controls in order to draw reliable and definitive science conclusions. In addition, it is very important to characterize terrestrial organic matter that gets sent to Mars on a space mission, because there is a distinct possibility that it would mix with the atmospheric and mineral samples on Mars and thereby skew the search for organic molecules and life on that planet.

GC-MS has been used to develop a library of spectral data related to rover materials and their composition [18,28,29]. By analyzing the GC-MS data for these broken-down components, it is possible to determine the exact makeup of the molecules that make up the rover components. In such a way, it is then possible to compare this spectral data with readings from *in situ* Mars sample analyses.

6.3 APPLIED SPECTROSCOPY

6.3.1 Laser Optogalvanic Spectroscopy and Discharge Plasmas

6.3.1.1 Overview

The laser-induced optogalvanic (OG) effect is a modus operandi to study atomic species in a discharge plasma by using a tunable laser, appropriate electronics, and data acquisition systems (as illustrated in Figure 6.19) [30–32]. The OG signal is produced due to the change in impedance of the electrical circuit when light of a well-defined frequency impinges on the plasma inside the discharge tube. Intense work has been done by the Misra and Han research groups [30–32] to build extensive spectral libraries of OG transitions involving neon in the visible region (590–670 nm) and argon in the UV (300–325 nm). To gain insight into the shapes and

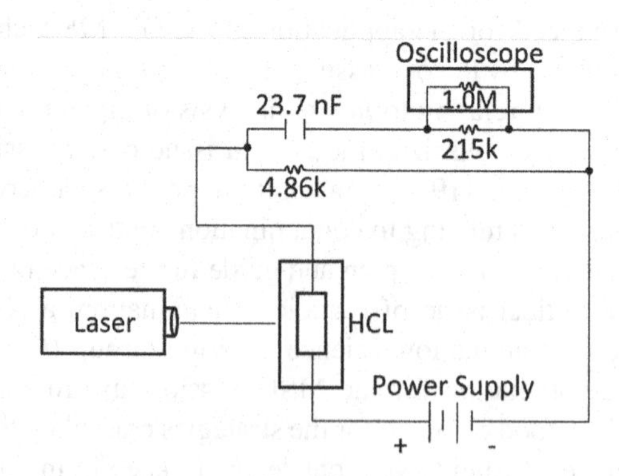

FIGURE 6.19 Experimental circuit used for laser optogalvanic spectroscopy [31].

characteristics of the OG waveforms, we have developed a model and employed a least-squares Monte Carlo subroutine fitting algorithm that is described later in this section. Specifically, we will illustrate the Monte Carlo approach for the $1s^22p^2$ transition of neon (at 659.9 nm) for discharge current values in the range 2–19 mA.

In a typical laser OG experiment, a laser beam (~ pulse width 5 ns) at the desired wavelength (659.9 nm) enters the so-called 'laser galvatron', which is a hollow cathode discharge lamp containing gaseous neon. The galvatron is part of an RC circuit in which the current can be limited and includes a power supply that regulates the voltage and supplies current in the required range (2–19 mA). The pulsed laser, when it interacts with the plasma, modifies the discharge current and deviates it from the steady state value and generates the OG signal voltage, which is averaged over 256 laser pulses and stored in a digital oscilloscope (Tektronix Model TDS 224; input impedance = 1 MΩ). The OG data is converted into ASCII format (or DOS text) and then used as input to the least-squares fitting routine cited earlier.

The key elements that were pursued relating to laser OG spectroscopy of rare gases (with special focus here on *neon*) in a hollow cathode discharge can be enumerated as follows: (1) the raw OG data were recorded for analysis and interpretation, together with the individual waveforms; (2) identification of all possible processes that contribute to the OG signal within the plasma of the sustained discharge and their incorporation into a comprehensive mathematical model that lends itself to continuous improvements and refinements of the associated rate equations; (3) the fitting of the time-resolved OG waveforms using a non-linear least-squares algorithm to determine the corresponding amplitudes, rates of signal decay, and the time constants for the circuit; (4) using the Monte Carlo least-squares technique in tandem with the developed rate equations to optimally model and simulate the OG waveforms with the required fitting parameters.

The neon atoms become ionized and interact with the incident laser beam and give rise to the OG signal voltage (as shown in Figure 6.19). The OG signal intensity can be understood in terms of the phenomena that occur when the laser frequency is in resonance and mimics the energy level separation of the neon species in the plasma, which in turn can be explained in detail by incorporating contributions from the following processes involving neon: (1) excitation due to collisions with electrons present in the discharge; (2) radiative depopulation of energy states; and (3) ionization due to electron collisions.

We will provide here a brief overview of the key processes that govern the generation of time dependent OG waveforms and the salient features of the mathematical rate equation model used to explain and simulate the shapes of the waveforms. The details have been provided in a series of publications from the Misra and Han research groups [30–32].

As illustrated in Figure 6.20, initially the neon atoms get excited by the laser from energy level L_1 to L_2. The excited neon atoms undergo a series of regular collisions and in turn get transferred from L_2 to L_2', following which there is a radiative decay and redistribution of the neon population in lower energy levels: L_3, L_4, and L_5. Because of this population dispersal, the total ionization rate is modified, and a discharge current flows in the OG circuit depicted in Figure 6.19.

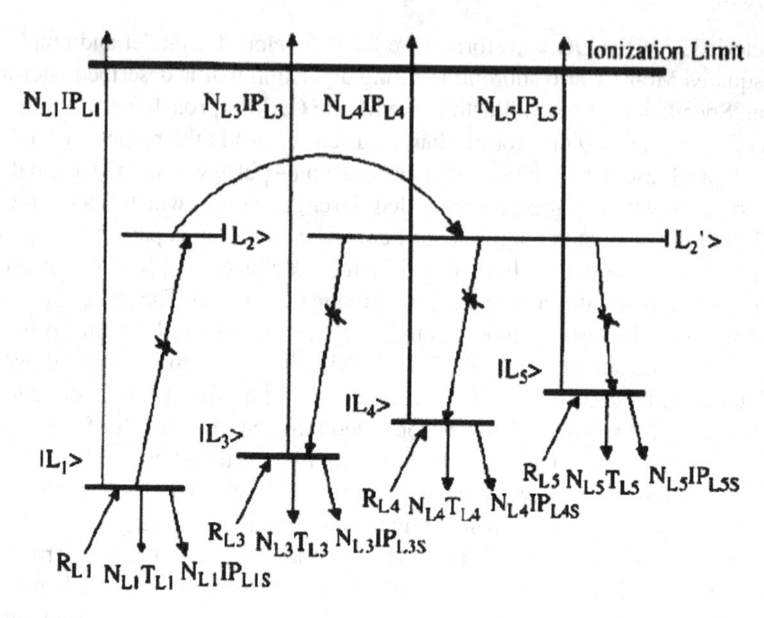

FIGURE 6.20 Simplified schematic energy-level diagram showing the optogalvanic transitions and the associated processes in neon. (Reprinted from Ref. 31, Copyright (2005), with permission from Elsevier.)

6.3.1.2 Case Study

In order to fix our ideas, let us consider a specific case study involving the 1s-2p OG transition in neon. We use the so-called Paschen notation to label the states involved in the explanation that follows. For the four 1s states and ten 2p states available, there are potentially a total of thirty allowed transitions that can take place radiatively. It should be kept in mind that of the four 1s states, two of them (1s2 and 1s4) decay very rapidly (~ ns), while the remaining two (1s3 and 1s5) are metastable and relatively long-lived. The focus in the present study is on the 1s2-2p2 OG transition observed at 659.9 nm.

Equation (6.17) was used to fit the shape of the recorded time-resolved OG waveforms for the 1s2-2p2 transition of neon:

$$s(t) = \sum_{j=1}^{j_{max}} \frac{a_j}{1 - b_j \tau} \left[e^{(-b_j t)} - e^{\left(-\frac{t}{\tau}\right)} \right] \tag{6.17}$$

In Equation (6.17), a_j are the amplitudes, b_j are the rate coefficients, and τ is the instrumental time constant that incorporates the response of the OG circuit and the plasma relaxation characteristics of the discharge that impact the formation of the waveforms.

As we can see from Figure 6.21, the agreement between the experimental and theoretically fitted waveforms is very good using both two (j = 1,2) and three (j = 1,2,3) terms in Equation (6.17).

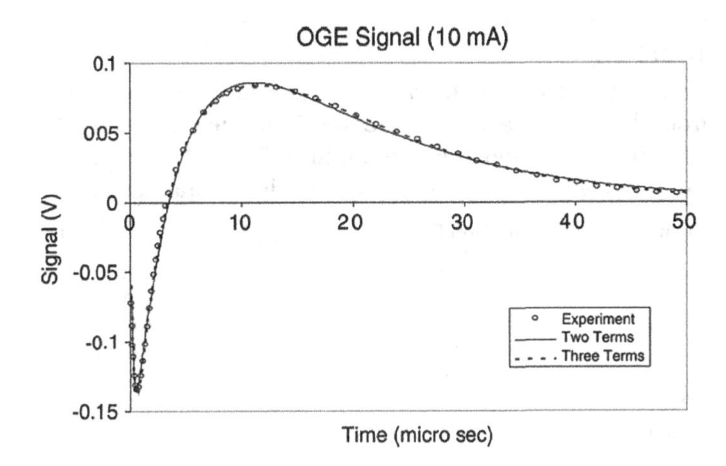

FIGURE 6.21 Overlayed comparison of the experimentally recorded OG waveform of neon with the theoretically fitted shapes obtained using both two and three terms in the rate equation model. (Reprinted from Ref. 31, Copyright (2005), with permission from Elsevier.)

The Monte Carlo approach employed here with three simultaneous terms has no issues with stability, despite strongly correlated fitting parameters, in contrast to previous conventional non-linear least-squares algorithm approaches that only allowed two-term fittings at a particular instant in Equation (6.17); otherwise, the parameters tended to blow up. There are some interesting aspects to the shape of the observed OG waveforms depending on whether a metastable state is involved or not. For example, the OG waveform shown in Figure 6.21, which represents the 1s2-2p2 transition of neon at 659.9 nm, is initially negative, but within a microsecond rises steeply and crosses the time axis around 3.45 ms, acquires a positive voltage value of about 0.08 V, and subsequently decays rapidly to zero. On the flip side, the 1s4-2p6 OG transition of neon at 630.2 nm, also recorded at the same current (10 mA) has an initial positive voltage and rapidly (within 5 ms or so) goes over to the negative side and subsequently gradually decays to zero. The primary difference between the two different waveform scenarios is that the latter (1s4-2p6) OG transition involves the initial metastable state (1s4), whereas the former (1s2-2p2) transition involves an initial state 1s2 that is not metastable.

It should be noted (cf. Table 6.7 and Figure 6.22) that although the rate coefficients b_2 and b_3 exhibit little variation over the current range (2–19 mA) investigated, the least-squares fitting employing 3 terms is an improvement over the one with two terms. Indeed, a small deviation is observed between the experimental and fitted OG waveforms for two terms, while the one performed with three terms shows no discrepancy at all.

The results obtained for the various fitted parameters (a_j, b_j, τ) are summarized in Table 6.7 for both two terms and three terms in Equation 6.17 for the OG neon waveform observed at 659.9 nm for a discharge current of 10 mA. For the same situation, Figure 6.22 provides a plot showing the variation of the rate parameters b_j ($j = 1, 2,3$) over the current range 2–19 mA.

TABLE 6.7

Amplitudes, Decay Rates and Time Constant Determined from the Non-Linear Least-Squares Fit of the Observed OG Waveform for Neon at 659.9 nm for a Current of 10 mA, using Two and Three Terms, Respectively, in Equation (6.17)

Parameter	Two-Term Fitted Values	Three-Term Fitted Values
$\tau(\mu s)$	0.206	0.274
$a_1(V)$	0.461	−0.113
$b_1(\mu s^{-1})$	0.217	0.647
$a_2(V)$	0.288	−0.980
$b_2(\mu s^{-1})$	0.0731	0.134
$a_3(V)$		0.894
$b_3(\mu s^{-1})$		0.0966

Source: Reprinted from Ref. 31, Copyright (2005), with permission from Elsevier.

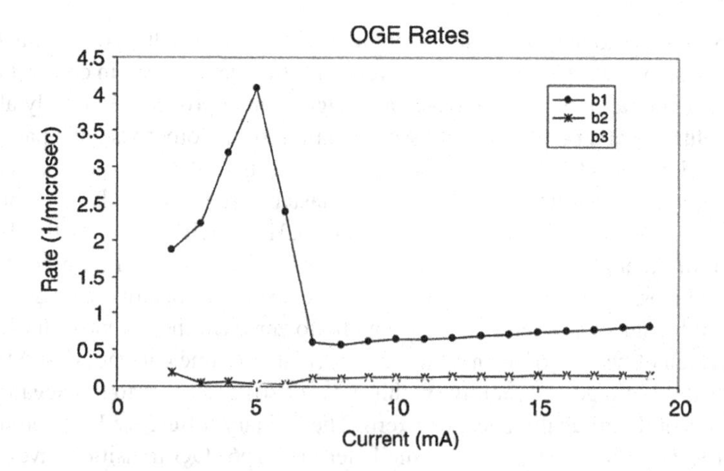

FIGURE 6.22 Graph illustrating average rates $b_j(j = 1, 2, 3)$ as a function of current (2–19 mA) for neon. (Reprinted from Ref. 31, Copyright (2005), with permission from Elsevier.)

It needs to be emphasized that the rate coefficients $b_j(j = 1, 2,3)$ collected in Table 6.8 and plotted in Figure 6.22 are the average of a pair of separately acquired data sets for the 659.9 nm OG neon transition recorded over the current range 2–19 mA.

The Han-Misra OG Monte Carlo technique designed to accomplish least-squares fitting of the OG signal waveforms is more stable than the standard non-linear least-squares fitting algorithm, despite the fitting parameters being extremely correlated. Consequently, it is anticipated that the Monte Carlo fitting algorithm will find wider applications in the fitting and modeling of experimental signal waveforms that involve exceedingly correlated parameters.

TABLE 6.8
Average Decay Rates $b_j (j = 1, 2, 3)$ Derived from the Non-Linear Least-Squares Fitting of the Observed OG Waveforms Corresponding to Two Separate Data Sets in the Current Range 2–19 mA

Current (mA)	$b_1 (\mu s^{-1})$	$b_2 (\mu s^{-1})$	$b_3 (\mu s^{-1})$
2	1.874	0.196	
3	2.227	0.0446	
4	3.192	0.0609	
5	4.075	0.0373	0.0316
6	2.392	0.0249	0.0199
7	0.592	0.111	0.087
8	0.57	0.116	0.0887
9	0.615	0.125	0.0934
10	0.648	0.133	0.0965
11	0.649	0.134	0.103
12	0.666	0.139	0.107
13	0.687	0.144	0.111
14	0.709	0.148	0.115
15	0.739	0.153	0.119
16	0.751	0.156	0.124
17	0.772	0.16	0.128
18	0.802	0.163	0.133
19	0.816	0.165	0.137

Source: Reprinted from Ref. 31, Copyright (2005), with permission from Elsevier.

6.3.2 Supersonic Jets and Laser-Induced Fluorescence Spectroscopy

6.3.2.1 Overview

Laser-induced fluorescence (LIF) spectroscopy is a versatile technique in which molecular species are irradiated with laser radiation in a specific wavelength range (normally in the ultraviolet for electronic spectroscopy) that is in resonance with the differences in molecular energy levels [33–39]. Such resonantly tuned radiation has a good probability of inducing a transition to the excited state of the molecule, which may be followed by relaxation of the molecules to the ground electronic state by spontaneous emission of a photon whose energy corresponds to the separation in molecular energy levels. Experimentally, the LIF technique involves scanning the wavelength of the probe laser (usually, a frequency-doubled tunable dye laser) and capturing the fluorescence intensity versus wavelength excitation spectrum.

An important advance associated with the LIF technique is to utilize a supersonic jet expansion to cool the precursor molecules subsequently followed by photodissociation of the precursor to produce ro-vibrationally cold gas phase free radical fragments. As a result of the low temperatures realized in the jet, collisional excitation and quenching phenomena are diminished, thereby the laser-induced fluorescence process can dominate, and simplified spectra of moderately sized radicals can be recorded and analyzed in detail [33].

It is well known that neutral free radicals play key roles as chemical intermediates in driving combustion and atmospheric processes in the gas phase. Several new techniques have been developed in the past several decades that have enabled greater understanding of these short-lived molecules. Among others, infrared (IR) and microwave (MW) absorption, resonantly enhanced multiphoton ionization (REMPI), and laser-induced fluorescence (LIF) spectroscopy technique have proved highly useful in characterizing such transient molecular species.

The alkoxy (RO, R=CH_3, C_2H_5, i-C_3H_7) radicals and the alkylthio (RS) radicals are important chemical intermediates in combustion reactions and in photochemical air pollution. It is well established that the combustion of organic fuels involves a complex sequence of reactions. Interestingly, most of the reactions proceed via the involvement of free radicals as chemical intermediates. It is envisioned that an elucidation of the spectroscopy and chemical kinetics of these chemical intermediates will provide insight into their participation in combustion reactions and the efficacy of organic fuels [34–39].

LIF spectroscopy, in conjunction with a supersonic jet expansion, has been used to study the RO radicals at low temperatures. In a typical LIF experiment, alkyl nitrite (RONO) molecules are photolyzed with an excimer laser (KrF at 248 nm) to generate RO fragments, which are subsequently excited from the ground state to excited state(s) by the absorption of dye laser radiation of specific frequencies. The energetically excited RO molecules can subsequently decay back to the ground state by spontaneous emission or alternately termed 'fluorescence' [33].

6.3.2.2 Experimental

A timing control circuit was designed to first send a pulse (excimer charge) to an excimer laser that orders it to begin charging its capacitor bank. A discharge of the capacitors results in the energy needed to fire the excimer laser, which is used to photolyze a suitable precursor and produce the appropriate free radical, e.g., ethyl nitrite (C_2H_5ONO) is photodissociated at 193 or 248 nm to generate the ethoxy (C_2H_5O) radical and nitric oxide (NO). Alkylthio (RS) radicals were produced by photolysis of either dialkyl sulfide (R_2S) or dialkyl disulfide (R_2S_2). Each organic precursor was allowed to flow with high pressure (8–10 atm) helium gas and entered a vacuum chamber through a 0.5 mm pulsed nozzle. Once the excimer laser fires, after a delay determined by the timing control circuit, a second trigger pulse orders the Nd:YAG laser to fire, and the resulting laser pulse then causes the dye laser to fire at a suitably tuned wavelength. A characteristic separation of about 10–12 mm was kept between the photolysis and probe lasers. The optical laser beam emanating from the dye laser is frequency-doubled using a non-linear crystal (e.g., KDP) to produce a UV pulse that excites the free radical. The molecular radical subsequently decays to the ground electronic state via spontaneous photon emission. An appropriate focusing lens is used to capture the laser-induced fluorescence photons onto a photomultiplier tube, which converts the optical signal into an electric current. A filter in front of the photomultiplier tube minimizes the scattered light entering it. The output from the photomultiplier tube is sampled by a boxcar integrator that produces a DC level, which is a measure of the LIF intensity. The timing control circuit used provides the trigger pulse to the boxcar and allows the synchronization of the boxcar gate with the fluorescence signal. The analog signal from the boxcar output is converted to digital and relayed to a personal computer for processing. A scan of the dye laser wavelength yielded the excitation spectrum of the free radical. The linewidth of the tunable dye laser pulse was about 0.2 cm^{-1} and could be enhanced to 0.07 cm^{-1}. Medium resolution LIF spectra were recorded and utilized for vibronic assignments and higher resolution spectra used for rotational assignments. Frequency calibration of the excitation spectra was done by recording concomitantly either the absorption spectrum of iodine or the OG spectral data of neon and argon.

6.3.2.3 Case Study

As a case study, we will consider alkoxy (RO) radicals in some detail, and methoxy (CH_3O) in particular. CH_3O has an orbitally degenerate ground electronic state (X^2E) and is therefore expected to exhibit Jahn-Teller distortion. A first order Jahn-Teller distortion leads to spontaneous breaking of the nominal C_{3v} symmetry in CH_3O. CH_3O is an intriguing free radical which possesses a quenched electronic angular momentum, and the associated reduction in spin-orbit splitting gives rise to striking features in the $A^2A_1 - X^2E$ spectral transitions. The spin-orbit interaction in CH_3O splits the X^2E electronic level into two components: $^2E_{1/2}$ and $^2E_{3/2}$. As a result, any vibronic level belonging to the A^2A_1 state can have possible transitions to both spin-orbit components of the X^2E state [33–37].

The RO radicals were produced *in situ* in the supersonic expansion by excimer laser (KrF at 248 nm) photolysis of freshly synthesized RONO. Jet-cooled RS radicals were produced by the photodissociation of commercially available precursors R_2S_2 with excimer laser pulses at 248 nm. The photolysis excimer pulse energy was typically 70–80 mJ. An Nd:YAGpumped dye laser was frequency-doubled to obtain the excitation spectra of CH_3O and C_2H_5O, whereas an excimer-pumped dye laser was employed to record the LIF spectra of i-C_3H_7O and the RS (R=CH_3, C_2H_5, i-C_3H_7) radicals. Typically, the dye laser pulse energy in the UV was about 1–3 mJ.

Fluorescence from the excited state of RO was wavelength-resolved using a 0.6 m monochromator (~0.3 nm resolution) [37]. Several vibrational frequencies have been obtained for the alkoxy molecules in the ground state, and the experimental uncertainty in the vibrational assignments is estimated to be about 20 cm^{-1}. Dispersed emission spectra for CH_3O and CH_3S were recorded by exciting the radicals at the wavenumber of a particular transition near a specific band maximum. The LIF photons were focused onto the entrance slit of a 0.6 m monochromator (Jobin Yvon HRS 2), with 0.2 mm wide slits and a resolution of 0.3 nm. The output signal was monitored at the exit slit and gathered by a photomultiplier tube and subsequently transferred to a microcomputer-driven data procurement system. Both the photolysis and probe lasers were run at 10 Hz, and each data point was an average of 10 measurements. The monochromator grating was scanned at 24 Å per minute and calibrated with reference mercury lines using a 0.025 mm wide entrance slit. A least-squares fit of the data provided an equation for the calibrated wavenumbers versus the reading of the monochromator dial.

Figure 6.23 is an illustrative dispersed fluorescence scan for the 2_0^1 band of CH_3O. The first doublet on the left of the scan corresponds to a mix of the pump laser wavelength and the emission produced as a result of the CH_3O transition from the

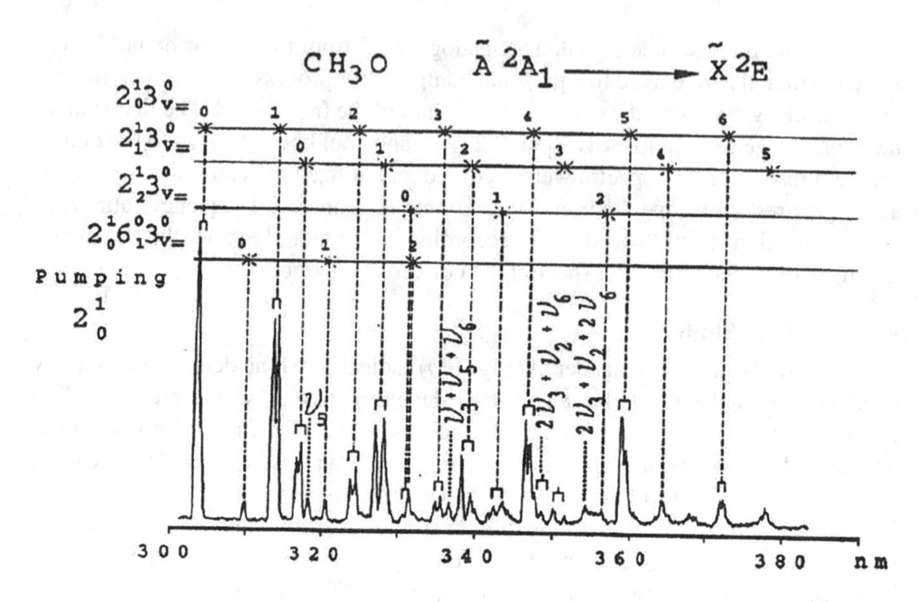

FIGURE 6.23 Laser-excited wavelength-resolved emission spectrum of CH_3O obtained when the 2_0^1 band was pumped [37]. A helium carrier gas pressure of 200 psi was maintained behind the nozzle and the time delay between the photolysis and probe lasers was kept constant at 6 μs.

vibrational level in the excited $^2A_1$ state to levels in the lower spin-orbit split $^2E_{1/2}$ and $^2E_{3/2}$ states. The second doublet in Figure 6.23 shows a pair of transitions separated by 61 cm^{-1} (akin to the first 65 cm^{-1} wide doublet feature), and these doublets are characteristic of the spin-orbit splitting of methoxy in the ground electronic state. Several v_3 progressions are visible in Figure 6.23 and have been assigned in conjunction with v_2, v_5, and v_6 vibrations. A thorough analysis of the wavelength-resolved emission spectra recorded for CH_3O provided the following vibrational frequencies for the X^2E state: v_1'' =2953 cm^{-1}, v_2'' = 1375 cm^{-1}, v_3'' = 1062 cm^{-1}, v_4'' =2869 cm^{-1}, v_5'' = 1528 cm^{-1} and v_6'' = 688 cm^{-1}.

As is well known, the methoxy radical is a free radical generated in the troposphere via the breakdown of methane by the hydroxyl radical. It subsequently reacts with molecular oxygen to generate formaldehyde as a stable product. The actual chemical reaction involving methoxy (CH_3O) production in the troposphere due to the breakdown of methane (CH_4) by the hydroxyl (OH) radical occurs via the following sequence of reactions [38,39]:

$$OH + CH_4 \rightarrow CH_3 + H_2O \tag{6.18}$$

$$CH_3 + O_2 + M \rightarrow CH_3O_2 + M \left(\text{where M is a third body}\right) \tag{6.19}$$

$$CH_3O_2 + NO \rightarrow CH_3O + NO_2 \tag{6.20}$$

The CH_3O radical generated reacts with oxygen (O_2) to produce formaldehyde (HCHO) as a stable product:

$$CH_3O + O_2 \rightarrow HCHO + HO_2 \quad (6.21)$$

The reaction of CH_3O with O_2, resulting in HO_2 and HCHO, is considered a significant atmospheric removal mechanism for CH_3O. This reaction is also important in the oxidation of hydrocarbons in flames. It is the dominant channel for the oxidation of CH_3O by O_2. The series of reactions involving oxidation of methane has significant ramifications for the overall ozone budget and anthropogenically induced perturbations in its concentration, since NO is oxidized to NO_2 by both CH_3O_2 and the product radical HO_2. In turn, this is followed by the reactions:

$$NO_2{}^{hv} \rightarrow NO + O \quad (6.22)$$

$$O + O_2{}^{M} \rightarrow O_3 \quad (6.23)$$

The chemical kinetics of the methoxy radical can be studied effectively using the LIF technique [38,39]. Different carrier gases (e.g., He, Ar, N_2) can be used to entrain the vapors of the methylnitrite precursor and transport it to the photolysis region. In a two-step process, first the methylnitrite is photodissociated by an excimer laser (at 248 or 193 nm) to generate the methoxy radical, and then its subsequent excitation by a tunable frequency-doubled Nd:YAG-pumped dye laser. The methoxy radical was reacted with molecular oxygen (maintained at pressures of 0–40 torr) over a range of temperatures (21.5–150°C), and the chemical kinetics of CH_3O were studied by keeping tab of the LIF signal over time.

Specifically, the decay of the methoxy fluorescence signal was recorded, while the two lasers (photolysis and probe) were delayed in time. An accurate determination of the rate constant for the reaction of CH_3O with O_2 was possible by making Stern-Volmer graphs of the inverse first-order decay constant (τ^{-1}) as a function of the O_2 pressure over the temperature range 21.5–150°C.

Rate constants were obtained as a function of temperature in the range 21.5–150.0°C for the gas phase reaction of the methoxy (CH_3O) radical with molecular oxygen (O_2) and an appropriate Arrhenius expression $k(T) = 5.48 \times 10^{-13} \exp[-2181/T(K)]$ cm^3 $molecule^{-1}$ s^{-1} derived.

6.3.3 Laser-Induced Breakdown Spectroscopy

6.3.3.1 Overview

The technique of laser-induced breakdown spectroscopy (LIBS) utilizes powerful lasers to impinge on the focused surface area of a target sample to vaporize it and generate a plasma containing a mix of ions, electrons, atoms, and molecular fragments. The excited species in turn emit characteristic radiation that can be collected using either a monochromator or a spectrograph equipped with a CCD camera and

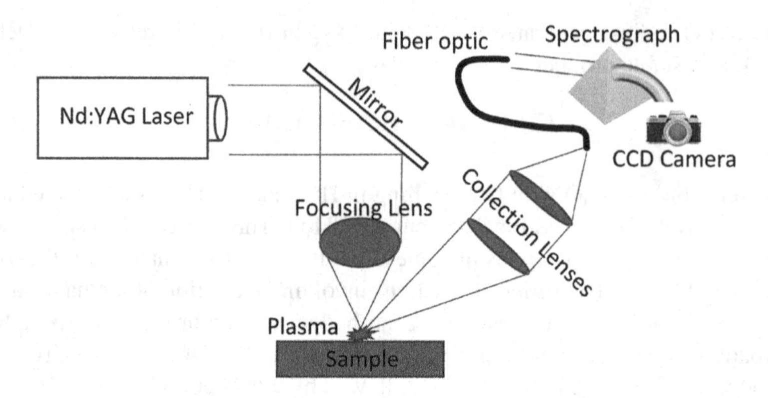

FIGURE 6.24 Schematic of a laser induced breakdown spectroscopy (LIBS) optical system [40].

subsequently processed to generate a spectrum [40]. The targeted sample can be in the solid, liquid, or gas phase.

In the LIBS experimental arrangement shown in Figure 6.24, a pulsed Nd:YAG laser produces energetic pulses that are directed with a mirror and focused on the sample with a converging lens and the resulting plasma emission is gathered by an optical fiber and detected with a spectrograph-CCD camera system.

The availability of short-pulsed (~ picoseconds [ps] and femtoseconds [fs]), high-powered lasers has opened new avenues for LIBS-driven research. In the traditional LIBS approach, nanosecond pulses are used to vaporize a sample, which implies lower power for the same laser pulse energy as compared to ps or fs pulses. By using short-pulsed ps and fs laser pulses, the available laser energy is deposited on the sample in orders of magnitude shorter time interval leading to higher power density and intensity of laser irradiation of the targeted sample.

6.3.3.2 Femtosecond vs. Nanosecond LIBS

As a LIBS case study [41–42], we compare the effects of a nanosecond-pulsed laser with a femtosecond one when they are used to irradiate a sample of solid brass (see Figure 6.25). The figure clearly shows a cleaner ablation realized with 40 pulses of a 100 fs laser as compared to a 4 ns laser. On the left of Figure 6.25, one can see cleaner crisp craters with much less debris formed using the 100 fs laser, while in contrast on the right the 4 ns laser shows an elevated rim around the edges of the crater symptomatic of local melting of the brass.

In Figure 6.26, we compare the effects of ps and fs pulses on copper and aluminum samples. It is clear from the illustration that the shorter the laser pulse, the greater the chance of all the associated energy to be deposited into a smaller volume of the target material, prior to its thermal expansion; therefore, there is a crisper and cleaner removal of the material via ablation. In the case study presented here, the fs laser pulse produces a lower rim (top row, Figure 6.26) in the crater, as compared to the ps laser (bottom row, Figure 6.26). This happens because the longer pulsed ps

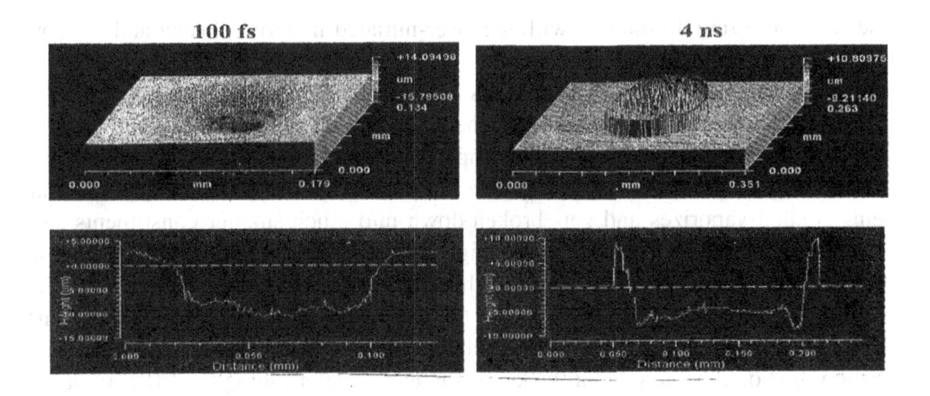

FIGURE 6.25 Crater profiles compared in brass samples after 40 pulses using a 100 fs and 4 ns laser. (Reproduced with permission of the Cambridge University Press through PLSclear, Laser Induced Breakdown Spectroscopy by A. Miziolek, V. Palleschi, and I. Schechter, Copyright (2006).)

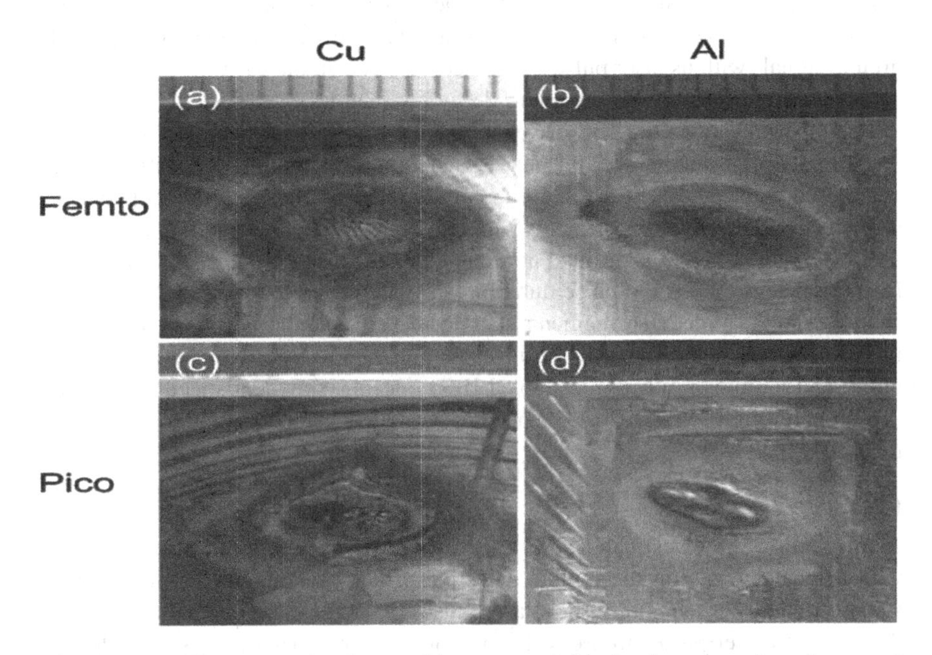

FIGURE 6.26 Femtosecond pulses on (a) copper and (b) aluminum samples; picosecond pulses on (c) copper and (d) aluminum samples. The scales appearing on the tops of the photographs are in mm. All four photographs here are presented on the same scale. (Republished with permission of the Royal Society of Chemistry, from Remote LIBS with ultrashort pulses: characteristics in picosecond and femtosecond. P. Rohwetter et al. [43], *J. Anal. At. Spectrom.* 19, 437–444 (2004); permission conveyed through Copyright Clearance Center, Inc.)

laser has more time to interact with the laser-initiated material plasma and thereby raises its temperature; whereas, the shorter duration fs laser pulse gives rise to a shorter lifetime plasma and therefore less intense background radiation as well.

Let us compare in some detail the characteristics of the LIBS material plasma when produced by ns laser pulses as compared to fs pulses. Specifically, the plasma generated by ns pulses is referred to as a hot plasma, whereby the sample material being studied vaporizes and gets broken down into much smaller constituents and therefore accurate identification of the original sample poses a problem. Besides this difficulty, a hot plasma gives rise to a white light continuum that produces a background signal, which in turn may swamp the low intensity emission emanating from the breakdown LIBS fragments and thereby interfere with their detection and identification. It should be noted that the key parameter in determining the LIBS-initiated sample breakdown is governed by the irradiance or intensity of the laser pulse, which depends on the laser beam size hitting the target, pulse energy and duration. The ns pulses of lower intensity hitting a target generate a plasma, whereas the shorter fs pulses of higher laser intensity are able to fragment the target material into smaller pieces down to atoms.

As a result, with fs laser pulses, one is able to collect the desired atomic line signals of the targeted species above the background continuum with measurable signal-to-noise (S/N) ratio. In addition, it is good to keep in mind that the high background signal (with ns laser pulses) could cover up desirable early signals produced by excited state species with ultrashort decay lifetimes.

An important aspect of LIBS-driven investigations is the knowledge that subnanosecond pulses produce cooler plasma, but an enhanced contrast S/N ratio must be considered with caution, with the duration of the laser pulse an important parameter to keep in mind [40]. In experiments with potassium nitrate and sodium nitrate samples, we noted a high background at around 750 fs pulse duration for well-defined laser intensities. When the pulse duration was lengthened (>750 fs and ~1 ps), the signals from the molecular fragments start decreasing, and those from atoms begin to flourish. It was noted that for the investigations associated with both potassium nitrate and sodium nitrate, the desirable LIBS molecular fragment-specific signatures had pronounced S/N ratios for enhanced laser intensities and pulse durations in the range 100–700 fs.

6.3.3.3 Case Studies

In this section, we investigate additional case studies involving the LIBS technique with a special focus on the remote detection of surrogate explosive molecules at stand-off distances using ultrashort laser pulses [41]. The approach was to consider the use of fs laser pulses to break down relatively large-sized explosive molecules into small molecular fragments and radical species in the excited state, so that they can produce detectable emission at stand-off distances of the order of 10s to 100s of meters. Most explosive molecules generate emission in the mid-IR (3–10 μm) region of the electromagnetic spectrum, which would be a challenge to detect and quantify, since mid-IR detectors lack sensitivity over distances as large as 10s to 100s of meters. One way around this is to use fs laser pulses and produce smaller fragments (e.g., NO_2, NO, CH_3, and CN) of the large explosive molecules, excite these smaller pieces into higher electronic states, and induce higher overtones of vibration in these species. As a consequence, the ensuing fluorescence emanating from these smaller fragments would be in the visible and near-IR regions of the electromagnetic spectrum and therefore possess higher energy and be detectable by more sensitive detectors (e.g., intensified CCDs with an imaging spectrographs) available in this regime, as reported for spectra of species such as NO, NO^+, and NO_2 [41].

Another important characteristic of ultrashort fs pulses is that they can be focused both longitudinally and transversely while propagating in air by utilizing the nonlinear properties of the medium, which is an important consideration for transporting energy effectively over longer stand-off distances, in contrast to merely using transverse focusing optics by itself.

The traditional LIBS technique has been employed effectively for both civilian and military applications. In addition, the 2012 Mars Science Laboratory (MSL) mission sponsored by the National Aeronautics and Space Administration (NASA) has an instrument called 'ChemCam' aboard the Curiosity rover that facilitates LIBS measurements remotely on the surface of Mars. ChemCam is short for Chemistry and Camera complex, which houses a pair of instruments, one makes LIBS measurements on rocks and minerals at a distance of 7 m from the rover on the Mars surface, and the other comprises of a Remote Micro Imager (RMI) telescope for recording good quality images of such objects. The LIBS system is designed with a 1064 nm laser (15 mJ energy/pulse and 5 ns pulse width) capable of being focused to a sub-mm spot size and of resolving rocks and Mars terrain features of the ~1 mm at stand-off distances as long as 7 m. The Mars Exploration Program (MEP) is served well by miniaturization of such instruments, so that they are compact and portable and consume low power for long-term space missions designed to identify biosignatures, such as the recently launched Mars 2020 mission carrying the Perseverance rover that landed on the surface of Mars on February 18, 2021.

If conventional LIBS were to be replaced by ultrashort-LIBS, the system would be able to generate molecular signatures and provide improved detection capability and would thereby enable the acquired data to offer more meaningful information and conclusions about the possibility of life on Mars.

6.3.4 Nuclear Magnetic Resonance

Nuclear magnetic resonance (NMR) spectroscopy can be effectively used to elucidate the detailed structure of organic molecules and in the study of molecular physics for both crystalline and non-crystalline materials [44]. A typical NMR setup can be seen in Figure 6.27.

All nuclei are electrically charged, and numerous nuclei possess spin. Owing to their intrinsic spin, the nuclei behave like a magnet and have a magnetic moment $\boldsymbol{\mu}$. When an external magnetic field is applied to a sample, a spin-1/2 nucleus will give rise to a magnetic field opposing the external B-field at a higher energy than the base energy, while the nucleus with spin at lower energy generates a magnetic field in the same direction as $\mathbf{B}_{\text{external}}$. The corresponding energy gap is in the radiofrequency range. Therefore, nuclei present in a constant strong magnetic field can be perturbed by a weak oscillating magnetic field. At resonance, the weak oscillating magnetic

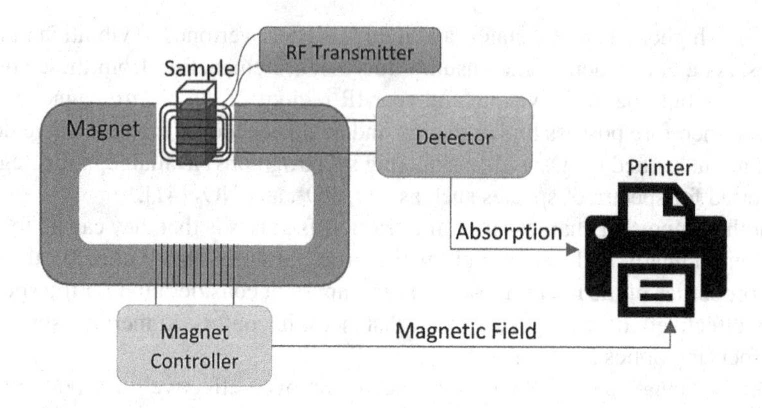

FIGURE 6.27 Schematic of a typical NMR device.

field matches the characteristic intrinsic frequency of the nuclei; that is, when the spin returns from the higher energy state to the base level, energy is emitted at the same radio frequency. The signal that corresponds to the energy transfer is measured to yield an NMR spectrum for the nucleus being studied.

The oscillating magnetic field in the form of a radiofrequency pulse applied orthogonal to the constant external **B**-field brings about a forced precession of the field at the so-called 'Larmor frequency', which results in a separation of energy states given by:

$$\Delta E = h\nu_0 \tag{6.24}$$

where h is Planck's constant and $\boldsymbol{\nu}_0$ is the frequency of electromagnetic radiation corresponding to the energy interval. Larmor precession describes the precession of the magnetic moment $\boldsymbol{\mu}$ of the nucleus about $\mathbf{B}_{\mathbf{ext}}$. The nuclear magnetic moment $\boldsymbol{\mu}$ experiences a torque $\boldsymbol{\tau}$ due to $\mathbf{B}_{\mathbf{ext}}$ given by:

$$\tau = \boldsymbol{\mu}\times\mathbf{B}_{\mathbf{ext}} = \gamma\, \mathbf{J}\times\mathbf{B}_{\mathbf{ext}} \tag{6.25}$$

where **J** is the angular momentum associated with the magnetic moment $\boldsymbol{\mu}$ and γ is the gyromagnetic ratio of the nucleus defined by $\boldsymbol{\mu} = \boldsymbol{\gamma}\mathbf{J}$. The angular momentum vector **J** precesses about the axis of $\mathbf{B}_{\mathbf{ext}}$ with a frequency called the Larmor angular frequency of magnitude ω given by:

$$\omega = \gamma\, \mathbf{B}_{\mathbf{ext}} \tag{6.26}$$

Here the gyromagnetic ratio $\boldsymbol{\gamma}$ is defined by the expression:

$$\gamma = \frac{qg}{2m} = \frac{-eg}{2m} \tag{6.27}$$

where $q = -e$ is the charge of the particle and m and g are the mass and g-factor of the precession system, respectively. Thus, the Larmor angular frequency ω and the gyromagnetic ratio γ are intimately linked to the g-factor of the nuclear system, which incorporates the properties of the nuclear spin angular momentum and the orbital angular momentum and their coupling. An important characteristic of the Larmor frequency in NMR spectroscopy is that it has no dependence on the angle between the vectors $\boldsymbol{\mu}$ and $\mathbf{B}_{ext}$, which reflects the fact that the rate of precession of the magnetic moment vector $\boldsymbol{\mu}$ has no correlation with the spatial orientation of the nuclear spin distribution.

NMR spectroscopy involves measurement of the electromagnetic signal that is observed when the applied radiofrequency pulse generates an observable voltage [44]. The signal is Fourier-transformed to obtain an NMR spectrum of the sample under study. The resonant frequency of the energy transition is governed by the effective magnetic field felt by the nucleus being studied. The chemical environment of the nucleus also has an impact on the electron shielding, which in turn affects the effective magnetic field at the nucleus. Consequently, the resonant frequency provides insight into the chemical environment of the nucleus being investigated. A more electronegative nucleus gives rise to a higher resonant frequency. In addition, bond strains and anisotropy influence the frequency shift as well. Typically, a standard, such as tetramethylsilane (TMS), is employed as a proton reference frequency ν_0, whereby the chemical shift is taken to be zero for TMS. The chemical shifts δ of other nuclei are stated with respect to TMS by using a parameter Ξ, such that:

$$\nu = \Xi \nu_{TMS} \tag{6.28}$$

$$\delta = \left(\nu - \nu_0\right)/\nu_0 \tag{6.29}$$

where ν is the absolute resonance frequency of the sample and ν_0 is the absolute resonance frequency of a reference material (in this case, TMS).

6.3.5 Biomedical Imaging

Medical imaging creates a visual representation of the interior of a human body to reveal internal structures for diagnosis and treatment of disease, which is a part of the new field of Medical Physics. Medical imaging equipment uses technology from the semiconductor industry (e.g., CMOS integrated circuit chips, image sensors, biosensors, microprocessors, digital signal processors, and lab-on-chip devices). Biological imaging uses radiology and employs X-ray radiography and magnetic resonance imaging (MRI), and among other techniques, positron emission tomography (PET) [45]. Fluoroscopy produces real-time images of internal structures of the body for performing image-guided procedures. MRI does not use ionizing radiation and is therefore considered less harmful than computed tomography (CT) and X-rays; although there are other health risks due to tissue heating from exposure to RF field and presence of implanted devices (e.g., pacemakers). Traditional MRI creates a 2D image of a thin slice of tissue, while modern MRI instruments produce images in the form of 3D blocks. Nuclear medicine uses both diagnostic imaging and treatment of disease (e.g., PET can be used to monitor rapidly growing tissue, like infections, metastasis, and tumors); modern scanners can integrate PET and make possible hyphenated techniques such as PET-CT and PET-MRI, that can optimize image reconstruction using positron imaging for non-invasive diagnosis and subsequent treatment. Volume rendering techniques enable CT, MRI and ultrasound scanning software to generate 3D images that are very useful in diagnosing and treating diseased tumors.

6.3.6 Photodynamic Therapy

Phospholipid bilayers in the form of membranes are able to form vesicles called 'liposomes' that are able to incorporate an interior aqueous spherical compartment [46–49]. The reason liposomes prove useful is that they mimic real-life cell membranes and natural vesicles and can be effectively utilized to study the impact of laser irradiation at a fundamental molecular level. Indeed, liposomes are highly promising vehicles for transporting and delivery of drugs and therapeutic agents to localized targeted sites. In fact, quantification of the efficiency of such delivery mechanisms can lead to improved understanding of fundamental energy transfer processes involving electronic and vibrational states at the molecular level. In addition, such investigations would also have ramifications for research and clinical diagnostics and treatment at the cellular level, together with providing options for photoinduced treatment of malignant tumors using the so-called photodynamic therapy (PDT) technique. In particular, specific types of lipid vesicles can be designed and synthesized for encapsulating photosensitive molecules mixed with therapeutic agents that can be released at targeted sites via irradiation and so-called 'laser heating'. We have performed a series of investigations using laser excitation of sulforhodamine dye-encapsulated liposomes [46–49] and measuring the dye released following a phase transition due to laser heating.

In addition, fluorescence lifetimes were determined for release of both solution phase and encapsulated sulforhodamine dye via time-correlated photon counting measurements. Two kinds of experiments were conducted; in one set of investigations, a continuous wave mode-locked frequency-doubled Nd:YAG laser (operating at 532 nm) was directly employed to stimulate the fluorescence, while in the second a Spectra Physics Model 3500 tunable dye laser with shorter pulse duration was used for inducing the fluorescence.

An ITT 4129 microchannel plate detector was used in tandem with a fast constant fraction discriminator (Model: Tennelec TC454). The response function of the system obtained by detecting light from a scattering solution was ~ 70 ps FWHM.

Fluorescence decays involving the liposome-dye complexes had to be fit to the sums of three exponentials. The data were corrected for the differential non-linearity of the time-to-amplitude converter (TAC). In order to effectively study the effects of polarization, the principal data were collected at a series of polarization angles perpendicular, parallel, and at 54.7° (magic angle), relative to the exciting polarization direction. The measurements and analysis provided a comparison of the percentages of the three lifetime components for different dye concentrations. It provided a quantitative estimate of the relative contributions from rather weak membrane-bound complexes (longest lifetime component), partial-quenching (intermediate lifetime), and full quenching (shortest lifetime) processes relating to the sulforhodamine dye molecules. The quenching is typical of Forster energy transfer and the lifetime data has implications for energy transfer and quenching. The use of selective polarization allowed us to study the effects of molecular reorientation in solution. The sample was excited with a vertically polarized laser and either the vertical or horizontally polarized emission was collected, together with data at the magic angle. When the collection of sulforhodamine chromophores is illuminated with laser light, an anisotropy occurs because only those molecules with a component of the absorption transition moment parallel to the polarization vector of the incident radiation will be excited. The effects of plane-polarized exciting radiation can be understood effectively in terms of the relative contributions from three lifetime components for different concentrations of sulforhodamine dye present within the liposomes.

One of the main applications of PDT is in the treatment of malignant tumors. The cancer drug (e.g., Taxol) is encapsulated in liposomes, along with an organic dye that fluoresces when impacted by a laser at the correct wavelength, to enable targeted release of the drug at the localized site of the cancerous growth and is treated in a confined manner, so that there is minimal impact on normal adjacent cell tissue. Another area of treatment with PDT is age-related macular degeneration (AMD), which can cause loss of vision and blindness. AMD seriously impacts the macula, which is the central portion of the retina of the eye, by making it thin over time. As a result, growth of blood vessels occurs below the retina, followed by fluid leakage and associated loss of vision. In order to prevent this from happening, an ophthalmologist can utilize PDT treatment by injecting a reagent in the vein of a patient's arm, which then accumulates in the abnormal blood vessels. A laser light is then shone on the reagent through a specialized contact lens to interact with the photosensitive drug and generate blood clots that seal off the harmful blood vessels and thereby limit further vision loss.

6.3.7 Nanomaterials and Sensors

6.3.7.1 Overview of Carbon Nanotubes, Graphene, and Metal Oxides

There are a variety of nanomaterials that can be used as sensors. We will consider carbon nanotubes, graphene, and metal oxides. Carbon nanotubes (CNTs) can be synthesized in two forms, namely single-walled carbon nanotubes (SWCNTs) and multi-walled carbon nanotubes (MWCNTs). SWCNTs are essentially sp2 hybridized C atoms shaped in the form of a hollow cylinder that have typical diameters ~ 1 nm and lengths in the range ~ μm to cm. MWCNTs have the same cylindrical structure and are composed of multiple concentric SWCNT layers nested together to form the composite nanostructure. An important property of CNTs is their high aspect ratio (i.e., ratio of length to width) which makes them model systems for testing theories associated with one-dimensional systems. In order to understand the behavior and physical properties of CNT, it is critical to understand the concept of chirality, which is a property that indicates that the mirror image of an object cannot be superimposed on the original by undergoing any series of translations and rotations. It is clear that a good choice for the axis of symmetry of a SWCNT is along its longitudinal direction, which in turn specifies the alignment of the hexagonal carbons along the wall of the cylindrical shape of the CNT. A powerful idea in this realm is that by folding the planar graphene lattice, one is able to obtain both possible achiral forms of SWCNTs, namely, the 'armchair' and 'zigzag' configurations, and in turn also account for the helicity of the remaining types of chirally feasible SWCNTs that do not possess mirror symmetry about the axis of the tube. It is good to know that the electrical characteristics of SWCNTs are governed exclusively by their geometry [7,10].

Interestingly, graphene, which is a carbon allotrope constituted of C atoms that are covalently bonded in the shape of a planar hexagonal lattice, can be considered as a building block for other allotropes of carbon. Various nanostructures can be obtained by rolling individual graphite planes into spheres, cylinders, and stacking them one above the other to form Buckyballs (0-dimensional), carbon nanotubes (1-dimensional), and graphite (3-dimensional). Keeping in mind that carbon (C) is positioned as a Group IV element in the periodic table of elements, its four valence electrons forming the 2s and 2p atomic orbitals blend into hybrid sp^2 orbitals, which causes the s-bonded electrons to be strongly bound and constrained on the graphene planar surface. On the other hand, the $2p_z$ electrons are weakly bound by p-bonds via clouds of electrons dispersed perpendicular to the graphene surface.

A myriad of applications have found their way in our everyday lives that involve graphitic nanomaterials (e.g., graphene and carbon nanotubes) such as storage devices, supercapacitors, environmental gas and biochemical sensors, nanoscale probes for therapeutic applications, and in targeted delivery of drugs for treatment of diseases [7,10].

In addition to these materials, metal oxides also have a wide range of sensor-based applications [6,11]. Promising research regarding modern sensors has been aimed at understanding the growth, characterization, and physical properties of a wide range of oxide materials, including tin dioxide (SnO_2), titanium dioxide (TiO_2), zinc oxide (ZnO), etc. Such investigations offer the possibility of exploring the interaction between the surfaces of metal oxides and gases, such as CO, CO_2, SO_2, and NO_x. Among the various oxides mentioned, SnO_2 has been investigated most extensively because of its wide bandgap n-type semiconductor characteristics and its tunable physicochemical properties. In addition, SnO_2 is very useful in the development and production of lithium-ion batteries, solar cells, catalysts, transistors, and gas sensors [6,11].

6.3.7.2 Spectroscopy of Carbon Nanotubes, Graphene, and Metal Oxides

We will consider in this section the spectral properties of graphitic nanomaterials and metal oxides that are of relevance to energy storage and gas sensing applications [6,7,10,11].

Resonance Raman spectroscopy is a powerful non-invasive method for the characterization of SWCNT and other nanomaterials, such as graphene and metal oxides. It is possible to obtain rich comprehensive Raman spectra from samples with minimal effort by way of sample preparation, in contrast to the more costly scanning probe techniques, e.g., atomic force microscopy and transmission electron microscopy (TEM). The thermal characterization of SWCNTs, graphene and metal oxides (WO_3 and SnO_2) using Raman spectroscopy proves useful for applications relating to energy storage and poisonous gas sensing [6,7,10,11].

It is important to understand the behavior of SWCNTs under conditions of elevated temperatures to gain insight regarding their thermal expansion and thermal conductivity properties by monitoring the variation of the G^+ band in the Raman spectra of SWCNTs recorded over a range of temperatures (from room temperature to 200°C) [10]. An elegant method has been developed to determine the purity of a bundled sample of SWCNTs for gas sensing applications. The technique is centered around monitoring the slope of the linear wavenumber variation of the Raman G^+ band of the SWCNTs sample as a function of laser power density, which turns out to be directly correlated with the amount of pure SWCNT content in each bundle and subsequently confirmed and calibrated via direct TEM of the CNT sample under study.

Let us go over in some detail the three linear regions of very distinct slopes that are typically observed in the Raman G^+ band wavenumber plotted as a function of laser power: (1) For the initial low laser power levels (~1–15 mW), the straight line plot is fairly flat, as there is little thermal heating of the CNT sample by the laser; (2) in the middle intermediate laser power range (~15–20 mW), there is a small increase in slope (corresponding to a Raman shift of about 0.5–1 cm^{-1} from the graphite value of 1592 cm^{-1}); and (3) in the higher laser power levels (~20–25 mW), the G^+ band feature is ablated (Raman shift ~2 cm^{-1}). It should be noted that bundles of SWCNTs of higher purity will be reflected by the extended horizontal portion of region 1 of the Raman G^+ band feature mimicking the behavior at low laser power levels. Highly pure SWCNT samples have lower content of amorphous carbon and metal catalytic particles. The correlation of thermal conductivity with the slope of the Raman G^+ band feature versus laser power levels is an important design consideration for energy storage device applications and can be directly obtained from Raman spectral measurements [10].

Besides carbon nanotubes, other nanomaterials such as graphene and graphene oxide also possess characteristics of large surface area, high mechanical rigidity, and good stable thermal conductivity over a range of temperatures, which make them good candidates for superior gas sensors. Another way to enhance the selectivity and detection limits of gas sensors is to consider functionalization of graphene, and thereby be able to adsorb a substantially greater number of desirable gas molecules on the surface at ultralow concentration levels (~ parts per million and even lower), together with fast response times and shorter recovery intervals for reuse of sensors.

Such distinctive properties of graphene-based sensors have advantages over semiconductor metal oxide-based gas sensors that often suffer from selectivity deficiencies in targeting particular toxic gas species, as well as requirements of increased power consumption and elevated temperatures.

Another variant of graphene and graphite that is gaining in popularity are graphene nanoplatelets (GnPs), which are now being commercially manufactured by the exfoliation of graphite. Typically, GnPs are composed of a few layers of graphene. A major advantage of GnPs, in comparison to graphene, is that they can be produced cost-effectively on a mass scale, and thereby are finding a range of applications in the industrial sector, such as gas sensors, supple electronics, and energy storage devices [7,11].

Next, we consider metal oxide-based nanomaterials. The Misra research group at Howard University has done extensive work on the Raman spectroscopy of such metal oxides by investigating the Raman shifts as a function of humidity, temperature, and their exposure to environmentally harmful gases by studying details of how gas molecules are adsorbed in the crystal lattice structure of the metal oxides [6,11]. Changes in conductivity due to chemisorption of oxygen can be measured and quantified in response to various external stimuli to design sensitive gas sensors for detecting targeted toxic gases (e.g., NO_x, CO_2, CO).

6.3.7.3 Applications of Graphitic Nanomaterials and Metal Oxides for Toxic Gas Sensing

The focus in this section is on the sp^2 carbonaceous graphitic species. As a result of having a distinctive C atom as the only type of atom making up the unit cell, these molecules serve as good quantum systems for the study of their vibrational and electronic energy levels. Such quantum mechanical studies have led to elucidation of electron-phonon interactions in solids, exciton-associated phenomena, and other esoteric effects, such as Kohn anomaly, which can then be applied to enhance understanding of other 2D materials. Indeed, the Raman spectroscopy of graphitic nanomaterials has led to greater insight into novel physics phenomena associated with new forms of graphene, such as graphene nanoplatelets [10].

The focus has been on the investigations of the various bands (G, D, and G') of graphene and graphite, and the potential information provided by each, which are useful for designing effective sensors for toxic gas sensing. The G-band peak (~1586 cm^{-1}), common to all sp^2 carbons, has been used widely in the determination of thermal conductivity and thermal expansion characteristics of single-walled carbon nanotubes. In addition, functionalized graphene nanoplatelets have been studied.

The Tunistra and Koenig relationship dictates the relative intensities of the main D and G Raman bands for the three graphitic nanomaterials (i.e., graphene, graphite, and graphene nanoplatelets) that have been studied in depth in the Misra research group. Besides Raman spectroscopy, scanning electron microscopy (SEM) imagery and dimensional analysis specially performed on the graphene nanoplatelet samples [11] confirmed the 3D form of the graphite sample and the 2D form of graphene. On the other hand, interestingly the graphene nanoplatelets (GnPs) displayed equally the 2D and 3D features.

Further studies are warranted on functionalized GnPs and their suitability for detecting specific toxic gases in the ambient environment.

Metal oxides are one of the extensively researched materials for gas-sensing applications, and owing to a change in their conductivity due to adsorption, they are well suited for detecting an array of combustible, reducing, and oxidizing gases. Metal oxides identified for gas sensing are selected based on their electronic band configurations. The Misra group (in particular, Raul Garcia-Sanchez and Hawazin Alghamdi) [7,11,48] and affiliated researchers have developed experimental protocols and theoretical modeling for investigating the behavior of metal oxide gas sensor materials. We have utilized Raman spectroscopy for energy level, vibration, and surface characterization, to understand the stretching, bonding, and other modes associated with the metal oxide samples following gas exposure. Metal oxides can effectively detect toxic gases (SO_2, NO_2, NO and NO_2-N_2O_4) by monitoring gas concentration, exposure time, temperature, and humidity changes associated with the Raman spectra of metal oxides exposed to the gas [50].

As a case study, we consider the response of tungsten oxide-based nanomaterials under specific external stimuli (e.g., effects of humidity and temperature and exposure to harmful toxic gases) under controlled conditions in the laboratory. Raman studies on tungsten oxide (WO_3) variants investigating thermal effects were performed for both increasing temperatures (room temperature to 200°C) and in the reverse direction of decreasing temperatures (200°C to room temperature) to see if any hysteresis (memory) effects were observed for the Raman peaks, along with the Raman shifts in wavenumber and intensity variations as well. Here we will focus on studies done with 780 nm laser excitation of monoclinic WO_3 deposited on a silicon substrate.

The key vibrational features of monoclinic WO_3 observed in the Raman spectra can be summarized as follows: O-W-O stretch (807 cm^{-1}), W-O stretch (716 cm^{-1}) and O-W-O bend (271 cm^{-1}), respectively. Additional spectral features seen in the Raman spectra indicated the presence of moisture (water vapor) in the ambient air and were observed around 1550 cm^{-1} and corresponded to OH-O and W-OH vibrations due to the impact of humidity on the WO_3 sample.

We have observed the following key vibrational features of WO_3: W-O stretch (~808 cm^{-1}), W-O bend (~714 cm^{-1}) and W-O-W deformation (~275 cm^{-1}), respectively. A variety of parameters, namely, position, height, and width, of the distinct spectral peaks and their changes when exposed to toxic gas molecules have been documented [50]. In addition to experimental investigations, we have employed COMSOL modeling and LAMMPS simulations to understand better the interaction of the adsorbed gas molecules with the material surface using a variety of parameters to compare with experimental results and study the behavior of sensors even under extreme conditions of operation that may not be feasible in the laboratory. Such computational studies will facilitate the design of gas sensors for *in situ* field measurements.

This research, in turn, helps extend the body of knowledge regarding the behavior of metal oxide-gas sensing interactions and potentially helps develop better theoretical models for improved understanding of metal oxide-gas relationships, which in turn will make the future development of more effective gas sensors feasible [50].

6.3.7.4 Other Applications of Nanomaterials

Along with graphitic nanomaterials, metal oxides have been integrated in various fields [6,7,10,11]. For instance, metal oxides have been incorporated in research related to materials science, geology, mechanical and electrical engineering, condensed matter physics, and inorganic chemistry. Such oxides have been engaged in various applications such as catalysis, medical applications (i.e., biosensor technology), environmental monitoring (i.e., water treatment, air pollutant mitigation, industrial and military waste purification), optronic, etc. Numerous semiconducting metal oxides have been used in energy storage and solar cell applications, such as SnO_2, TiO_2, ZnO, NiO_2, etc. Specifically, tin dioxide (SnO_2) has been investigated for a wide variety of applications. It has been tested in solar cells as antireflection coating and solar energy absorption coating, in rechargeable lithium-ion batteries, and most commonly in gas sensing devices. Transparent conducting oxides (TCOs) (e.g., SnO_2, In_2O_3, CdO, Ga_2O_3, ZnO, etc.) have been the center of attention for the past 20 years due to their interesting and multifaceted properties. They are included in many applications such as high-resolution displays and screen, i.e., liquid crystal display (LCD), high-definition television (HDTV), organic light-emitting diode (OLED); portable smart devices; smart windows; and thin film photovoltaics. One example of TCO materials is SnO_2, which has a large optical transparency and small electrical resistance. Even though SnO_2 is transparent in the visible range of the electromagnetic spectrum, it still has large reflectivity in the infrared region. Therefore, both SnO_2 and fluorine doped (SnO_2/F) are commonly applied as a coating material in energy saving windows, where they are effective in limiting loss of radiative heat due to their small thermal emissions in comparison to uncoated glass.

6.4 SUMMARY AND CONCLUDING REMARKS

This chapter provides a comprehensive survey of various spectroscopic techniques that are in use today, along with in-depth illustrative case studies of important applications at the forefront of research and development in cutting-edge areas of modern-day technology.

ACKNOWLEDGMENTS

Some of the case studies illustrated in this chapter were facilitated by funding from the National Science Foundation (NSF) via the Research Experiences for Undergraduates (REU) Site in Physics at Howard University (Award Nos. PHY-1358727 and PHY-1659224) which is gratefully acknowledged. In addition, allocation support via Grant No. TG-DMR190126 through Extreme Science and Engineering Discovery Environment (XSEDE) for the Molecular Dynamics simulations of tin dioxide and graphene nanoplatelets is acknowledged. In particular, the case study on Raman spectroscopy of tungsten trioxide at elevated temperatures was part of Christina Craig's REU 2015 summer project, while the graphene nanoplatelets and machine learning case study was a direct result of the research performed by REU summer interns, Lia Phillips and Benjamin Concepcion (REU 2019 cohort). The case studies relating to the spectroscopy of jet-cooled alkoxy radicals and the associated chemical kinetics studies, along with the FTIR spectroscopy investigations centered on adsorption of gases on metals, were part of the Ph.D. dissertation research performed by Dr. Abdullahi Nur, Dr. Michael King, and Dr. Edward Dowdye, respectively, in Prof. Misra's *Laser Spectroscopy Laboratory* at Howard University. The laser optogalvanic spectroscopy case study is based on the Ph.D. research of Dr. Helen Major and Dr. Ogungbemi Kayode. Specifically, the case study relating to carbon nanotubes and graphitic nanomaterials draws on Dr. Daniel Casimir's Ph.D. dissertation work, and that focused on tungsten trioxide was part of Dr. Raul Garcia-Sanchez's Ph.D. dissertation. The LIBS case study cited in this chapter is based on Dr. Tariq Ahmido's Ph.D. dissertation work co-supervised by Drs. Antonio Ting and

Prabhakar Misra Additional case studies presented here relating to tin dioxide and graphene nanoplatelets, respectively, draw on the Ph.D. research currently pursued by Hawazin Alghamdi and Olasunbo Farinre. All of the undergraduate research and Ph.D. dissertation research cited here have been performed under the guidance of Prof. Misra at Howard University in Washington, DC. The support from research associates and postdoctoral scholars, namely, Xinming Zhu, and Drs. Bruce Shih, Mark Dubinskii, and Chandran Haridas are also gratefully acknowledged.

REFERENCES

1. Mark F. Vitha. *Spectroscopy: Principles and Instrumentation*, John Wiley & Sons, Hoboken, NJ, 2019, ISBN: 978-1-119-43664-5.
2. E. Roland Menzel. *Laser Spectroscopy: Techniques and Applications*, Marcel Dekker, New York, NY, 1995, ISBN: 0-8247-9265-3.
3. Max Diem. *Introduction to Modern Vibrational Spectroscopy*, John Wiley & Sons, New York, NY, 1993, ISBN: 0-471-59584-5.
4. Otto S. Wolfbeis (Editor). *Fluorescence Methods and Applications: Spectroscopy, Imaging, and Probes*, Volume 1130, Annals of the New York Academy of Sciences, Boston, MA, 2008, ISBN: 978-1-57331-716-0.
5. A. Michael, P. Misra, A. Farah, and V. Kushawaha. Electronic emission due to collisions involving low energy CHO^+ and H^+ ions and CH_4 and N_2 molecules, *Journal of Physics B: Atomic, Molecular and Optical Physics*, 25, 2343–2350, 1992. http://dx.doi.org/10.1088/0953-4075/25/10/014
6. H. Alghamdi, B. Concepcion, S. Baliga, and P. Misra. Synthesis, Spectroscopic Characterization and Applications of Tin Dioxide. In: Mubarak N.M., Khalid M., Walvekar R., Numan A. (eds.), *Contemporary Nanomaterials in Material Engineering Applications. Engineering Materials.* Springer, Cham, 2021. https://doi.org/10.1007/978-3-030-62761-4_11
7. Daniel Casimir, Raul Garcia-Sanchez, Olasunbo Farinre, Lia Phillips, and Prabhakar Misra, Raman Spectroscopy and Molecular Dynamics Simulation Studies of Graphitic Nanomaterials. In: Vinod Tewary and Yong Zhang (eds.), *Modeling, Characterization and Production of Nanomaterials: Electronics, Photonics and Energy Applications,* 2nd ed., Elsevier Science, Woodhead Publishing Series in Electronic and Optical Materials Series, New York, NY, 2021. ISBN-10: 0128199059, ISBN-13: 9780128199053.
8. D.M. Bower, P. Misra, M. Peterson, M. Howard, T. Hewagama, N. Gorius, S. Li, T. Aslam, T.A. Livengood, A. McAdam, and J.R. Kolasinski, Paper# EPSC2020-427, Session TP12 – Open Lunar Science & Innovation, Europlanet Science Congress 2020 (EPSC2020), Virtual Meeting, September 21–October 9, 2020. https://meetingorganizer.copernicus.org/EPSC2020/session/38432
9. Prabhakar Misra and Edward H. Dowdye, Jr. Mid-infrared Spectroscopy of Molecular Species that Drive Significant Atmospheric Processes, *Proceedings of the International Conference on LASERS 2001*, V. Corcoran and T. Corcoran (eds.), STS Press, McLean, VA, 2002, pp. 386–393.
10. Daniel Casimir. *Investigation of Thermal Expansion Properties of Single Walled Carbon Nanotubes by Raman Spectroscopy and Molecular Dynamics Simulation*, Ph.D. Dissertation, Howard University, Washington, DC, 2015.
11. Prabhakar Misra, Hawazin Alghamdi, and Olasunbo Farinre. Spectroscopic Characterization and Molecular Dynamics Simulation of Tin Dioxide and Functionalized Graphene Nanoplatelets. Paper # 8225, Tech Science Press, *Proceedings of the*

International Conference on Computational & Experimental Engineering and Sciences (ICCES), Phuket, Thailand, January 6-10, 2021, S.N. Atluri and I. Vusanovic, Editors, ICCES 2021, MMS 97, pp. 1–15, Springer Nature, Switzerland, AG, 2021. https://doi.org/10.1007/978-3-030-64690-5_4

12. Fred W. McLafferty. A century of progress in molecular mass spectrometry. *Annual Review of Analytical Chemistry*, 4, 1–22, 2011. https://doi.org/10.1146/annurev-anchem-061010-114018
13. Gary L. Glish and Richard W. Vachet. The basics of mass spectrometry in the twenty-first century. *Nature Reviews, Drug Discovery*, 2, 140–150, 2003. doi: 10.1038/nrd1011
14. Ulrich Boesl. Time-of-flight mass spectrometry: Introduction to the basics. *Mass Spectrometry Reviews*, 36, 86–109, 2017. https://doi.org/10.1002/mas.21520
15. Igor V. Chernushevich, Alexander V. Loboda, and Bruce A. Thomson. An introduction to quadrupole-time-of-flight mass spectrometry. *Journal of Mass Spectrometry*, 36(8), 849–865, 2001. doi: 10.1002/jms207
16. Dirk Nolting, Robert Malek, and Alexander Makarov. Ion traps in modern mass spectrometry. *Mass Spectrometry Reviews*, 38(2), 150–168, 2019. https://doi.org/10.1002/mas.21549
17. Ashish Chauhan, Manish Kumar Goyal, and Priyanka Chauhan. GC-MS Technique and its analytical applications in science and technology. *Journal of Analytical and Bioanalytical Techniques*, 5, 222, 2014. doi:10.4172/2155-9872.1000222
18. P. Misra, R. Garcia, and P.R. Mahaffy. Gas Chromatography and Mass Spectrometry Measurements and Protocols for Database and Library Development Relating to Organic Species in Support of the Mars Science Laboratory. *Proceedings of the Astrobiology Science Conference (AbSciCon 2010)*, League City, TX, 2010.
19. P.R. Young and W.S. Slemp. The performance of selected polymeric materials exposed to low earth orbit. *Polymers Advanced Technologies*, 9(1), 20–23, 1998.
20. R. Rampini, L. Grizzaffi, and C. Lobascio. *Materialwissenschaft Und Werkstofftechnik*, 34(4), 359–364, 2003.
21. C.W. Chang. Comparing surface particle coverage predictions with image analysis measurements. *Proc. SPIE 6291, Optical Systems Degradation, Contamination, and Stray Light Effects, Measurements, and Control II*, 62910K, 7 September 2006.
22. K.T. Luey, R.M. Villahermosa, and D.J. Coleman. Optical properties of contaminant droplets. *Proc. SPIE 5526, Optical Systems Degradation, Contamination, and Stray Light Effects, Measurements, and Control*, 15 October 2004.
23. B.C. Christner, J.A. Mikucki, C.M. Foreman, J. Denson, and J.C. Priscu. Paper presented at the 3rd International Conference on Mars Polar Science and Exploration, Academic Press Inc. Elsevier Science, Lake Louise, Canada, Oct 13–17, 2003.
24. E. Grosjean and G.A. Logan. Incorporation of organic contaminants into geochemical samples and an assessment of potential sources: examples from Geoscience Australia's marine survey S282. *Organic Geochemistry*, 38, 853–869, 2007.
25. J. Eigenbrode, L.G. Benning, J. Maule, N. Wainwright, A. Steele, H.E.F. Amundsen, and the AMASE 2006 Team. A field-based cleaning protocol for sampling devices used in life-detection studies. *Astrobiology*, 9(5), 455–465, 2009.
26. P.R. Mahaffy, D. Beaty, M. Anderson, G. Aveni, J. Bada, S. Clemett, D. Des Marais, S. Douglas, J. Dworkin, R. Kern, D. Papanastassiou, F. Palluconi, J. Simmonds, A. Steele, H. Waite, and A. Zent. [Unpublished white paper], MEPAG Topical Analysis Reports, Mars Exploration Program Analysis Group, 2003.
27. D.A. Flory, J. Oró, and P.V. Fennessey. Organic contamination problems in the Viking molecular analysis experiment. *Origins of Life*, 5, 443–455, 1974.

28. R. Garcia, P. Misra, I. ten Kate, and P. Mahaffy. Database and Library Development of Organic Species Using Gas Chromatography and Mass Spectral Measurements in Support of Sample Analysis at Mars. *Proceedings of the NSBE Aerospace Systems Conference*, Los Angeles, CA, 2010.
29. P. Misra, R. Garcia, P. Mahaffy, J. Canham, and D. Jallice. SAM/MSL Contaminants Spectral Library, NASA Goddard Space Flight Center New Technology Report (NTR) #: GSC-16547-1, NASA Technology Transfer System (NTTS), February 2, 2012. Published as "SAM/MSL Terrestrial Background Spectral Library" in *NASA Software Tech Briefs*, September 2014, p. 23.
30. X.L. Han, M.C. Su, C. Haridass, and P. Misra, Collisional dynamics of the first excited states of neon in the 590-670 nm region using laser optogalvanic spectroscopy, *Journal of Molecular Structure*, 695-696, 155–162, 2004.
31. Prabhakar Misra, Isha Misra, and Xianming L. Han. Laser optogalvanic spectroscopy of neon at 659.9 nm in a discharge plasma and nonlinear least-squares fitting of associated waveforms. *Nonlinear Analysis*, 71, e661–e664, 2009. doi:10.1016/j.na.2008.11.086
32. X.L. Han, H. Chandran, and P. Misra. Collisional rate parameters for the $1s_4$ energy level of neon 638.3 nm and 650.7 nm transitions from the analyses of the time-dependent optogalvanic signals. *Journal of Atomic and Molecular Sciences* 1(2), 118–125, 2010. doi: 10.4208/jams.051509.071209a
33. P. Misra. Spectroscopic Characterization of Cold Radicals Using the Laser-Induced Fluorescence Technique. In V. Corcoran and T. Corcoran (eds), *Proceedings of the International Conference on LASERS 2001,* STS Press, McLean, VA, 2002, pp. 375–378.
34. Prabhakar Misra, Xinming Zhu, Hosie L. Bryant, and Mohammed M. Kamal. Rotationally-Resolved Excitation Spectroscopy of the Alkoxy and Alkylthio Radicals in a Supersonic Jet. Paper TJ.7, *Proc. Fifteenth International Conference on Lasers '92*, Houston, TX, 696–701, 1992.
35. Prabhakar Misra. Laser Excited Spectra of the Jet-Cooled Ethoxy Radical. *Proceedings of the International Conference on LASERS '95,* STS Press, McLean, VA, 1996, pp. 813–817.
36. Xinming Zhu, Mohammed M. Kamal, and Prabhakar Misra. Laser-induced excitation and dispersed fluorescence spectra of the ethoxy radical. *Pure and Applied Optics*, 5, 1021–1029, 1996.
37. Prabhakar Misra, Xinming Zhu, Ching-Yu Hsueh, and Mohammed M. Kamal. Wavelength-Resolved Emission Spectroscopy of the Alkoxy and Alkylthio Radicals in a Supersonic Jet. Paper TJ.8, *Proc. Fifteenth International Conference on Lasers '92*, Houston, TX, 702–705, 1992.
38. Abdullahi H. Nur, Xinming Zhu, Mohammed M. Kamal, Hosie L. Bryant, Michael King, and Prabhakar Misra. Fluorescence Lifetimes and Kinetics of the Methoxy Radical. *Proceedings of the International Conference on LASERS '94*, Society for Optical & Quantum Electronics, STS Press, 1995, pp. 532–536.
39. Prabhakar Misra, Xinming Zhu, and Abdullahi H. Nur. Chemical Kinetics of the Reaction of Methoxy with Oxygen. *Proceedings of the International Conference on LASERS '95,* STS Press, McLean, VA, 1996, pp. 830-835.
40. Tariq Ahmido. Remote Sensing of Explosive Surrogates Using Ultrashort Laser Induced Breakdown Spectroscopy, Ph.D. Dissertation, Howard University, Washington, DC, 2011.
41. T. Ahmido, A. Ting, and P. Misra. Femtosecond laser-induced breakdown spectroscopy of surface nitrate chemicals. *Applied Optics,* 52(13), 1 May 2013. http://dx.doi.org/10.1364/AO.52.003048.

42. A. Miziolek, V. Palleschi, and I. Schechter, *Laser Induced Breakdown Spectroscopy (LIBS) Fundamentals and Applications*, Cambridge University Press, New York, 2006.
43. P. Rohwetter, J. Yu, G. Mejean, K. Stelmaszczyk, E. Salmon, J. Kasparian, J. P. Wolf, and L. Woste. Remote LIBS with ultrashort pulses: characteristics in picosecond and femtosecond regimes, *Journal of Analytical Atomic Spectrometry,* 19, 437–444, 2004.
44. Harald Ginther. *NMR Spectroscopy: Basic Principles, Concepts and Applications in Chemistry*, 3rd ed., Wiley-VCH, Weinheim, Germany, 2013. ISBN: 978-3527330003.
45. Justine Wallyn, Nicolas Anton, Salman Akram, and Thierry F. Vandamme. Biomedical imaging: Principles, technologies, clinical aspects, contrast agents, limitations and future trends in nanomedicines. *Pharmaceutical Research,* 36, 78, 2019. doi: https://doi.org/10.1007/s11095-019-2608-5
46. D.L. Vander Meulen, Prabhakar Misra, J. Michael, M. Khoka, and K.G. Spears. Quantitative Analysis at the Molecular Level of Laser-Neural Tissue Interactions Using a Liposome Model System. *Proc. SPIE (Society of Photo-Optical Instrumentation Engineers)*, 1428, 91–98, 1991.
47. D.L. Vander Meulen, P. Misra, M. Khoka, J. Michael, and K.G. Spears. Photorelease of liposome contents by dye-mediated localized heating induced by picosecond or nanosecond laser excitation. *Biophysical Journal*, 59(2), 627a, 1991.
48. P. Misra, J. Michael, D.L. Vander Meulen, M. Khoka, and K.G. Spears. Laser-Induced Release of Organic Dyes from Liposomes. *Proceedings of CLEO (Conference on Lasers & Electro-Optics)*, 10, 78, 1991.
49. D.L. VanderMeulen, Prabhakar Misra, J. Michael, K.G. Spears, and M. Khoka. Laser mediated release of eye from liposomes. *Photochemistry and Photobiology*, 56(3), 325–332, 1992.
50. Raul Garcia-Sanchez. *Characterization of Metal Oxide Gas Sensor Materials Using Raman Spectroscopy and Computer Simulations*, Ph.D. Dissertation, Howard University, Washington, DC, 2016.

7 Computer Modelling and Simulation, Artificial Intelligence and Quantum Computing

7.1 INTRODUCTION

These introductory remarks are related to computer modelling and simulation, artificial intelligence, and quantum computing and their applications in focused areas spanning large databases, graphitic and metal oxide nanomaterials, quantum machine learning, and self-driving cars. The process of creation and manipulation of computer-driven numerical, graphical, or algorithmic depiction of physical phenomena and everyday systems is termed 'computer modelling and simulation' [1]. One can consider simulation to imitate reality by using a computer program to model a particular concept or process; one can in this way study in detail real-life systems and phenomena and at the same time predict and/or optimize certain processes that affect their behavior.

7.2 MATHEMATICAL AND COMPUTER MODELLING

7.2.1 Model Development

To build a valid and credible simulation model [1], a five-step process can be followed: (1) choose a system, (2) identify the problem, (3) design the problem, (4) collect and start processing data, and observe the performance, and (5) analyze, interpret the results, and if needed make predictions.

7.2.2 Model Visualization

Visualization is a vital component of modelling and simulation. Typically, there are four types of visualization for modelling and simulation protocols, namely, (1) conceptual and diagrammatic, (2) quantitative, (3) seek and find, and (4) pattern and flow. Conceptual and diagrammatic visualization embodies the conceptual design of the model and the associated flow of logic. The quantitative visualization incorporates both semi-static and static graphs and time-series plots and statistics connected with the data. Besides 2D and 3D representations of the quantitative data in the form of graphs and plots, animated scatter plots employing both coordinate axes are also used to show dynamic variation of the data over time. Seek and find visualization permits the user to vary parameters and explore potential relationships between the

DOI: 10.1201/9781003213468-7

input and output typically via a graphical user interface (GUI). Pattern and flow visualization depict a dynamic display for an active simulation in progress in real-time usually via animated graphics that provide the user the ability to view interim data patterns and adjust as needed.

7.2.3 Data Preparation

Data preparation comprises cleaning, organizing, and collecting the data sets for building the requisite computer model. It is an activity designed to transform raw, dissimilar, and disorganized data sets into a clean, clear, and ordered view of the data. Automating and operationalizing the data preparation helps to significantly enhance the efficiency of data processing for building the computer model. Use of data visualization and machine learning techniques helps to address data quality issues and reduce the time for data preparation.

7.3 SIMULATION TECHNIQUES

7.3.1 Large-Scale Atomic/Molecular Massively Parallel Simulator

Molecular dynamics (MD) simulations incorporated within the large-scale atomic/molecular massively parallel simulator (LAMMPS) code [2] is a powerful way to simulate physical phenomena associated with nanomaterials. In this classic technique, the trajectories of the molecules being studied are generated according to Newton's equations of motion. The basic principle of algorithm-driven operation of MD simulation can be divided into four steps: (1) The initial atomic-positions, temperature, time step, particle number, etc. of the simulation are provided to initialize the system being modelled. (2) The energy and forces of the model system are computed. The forces acting on each particle comprising the system due to its neighboring particles within a specified range of interaction are computed by calculating the gradient of the potential function. The potential function describes the interactions between the constituent particles and their neighbors up to a certain cut-off distance. The choice of potential used in defining the system being studied is critical for determining the accuracy and success of the MD simulation. (3) The particle trajectory consisting of positions and velocities of the particles comprising the molecular system—using discrete/successive time steps—is developed from the forces computed in step 2 and based on Newton's second law of motion. (4) The results and analyses of the trajectories of the modelled system are then used to predict its dynamic and thermodynamic properties.

To understand the interaction between a metal oxide sensor and a gas, different empirical interatomic potentials such as Lennard-Jones potential need to be studied. The Lennard-Jones potential by itself is not capable of modelling bonded interactions such as bond bending, bond breaking, torsion and angular potentials. To resolve this, a many-body potential such as the adaptive intermolecular reactive empirical bond order (AIREBO) potential is needed. This potential is used to model covalent bonding interactions in solid-phase hydrocarbons.

LAMMPS can be used to study the thermal properties of atoms and molecules, as well as energy-related variables; these include thermal conductivity and energy potential. In addition, LAMMPS can be used to study Monte Carlo experiments, many-body potentials, electrostatic potentials, and pairwise potentials [2]. LAMMPS has been used to model metal oxide–gas molecule interactions [3–6]. Figure 7.1 shows the structure of WO_3.

Figure 7.2 shows LAMMPS being used to model the conductivity of TiO_2 films [4]. As shown in the figure, LAMMPS can be utilized in the simulation of metal oxides. LAMMPS has also been used for modelling the electrical, thermal, and structural properties of carbon surfaces.

Metal oxide materials come in a variety of structural forms and lattice orientations. These different crystal structures and orientations for the same metal oxide influence the many properties of metal gas sensors, such as sensitivity and selectivity. These changes make metal oxide gas sensors dependent on the manufacturing process. As such, it becomes important to study these metal oxide materials at the atomic and molecular level, which makes Raman spectroscopy an ideal tool for characterizing the properties of the metal oxides used for these sensors [7–10]. It also becomes important to model these properties using molecular dynamics.

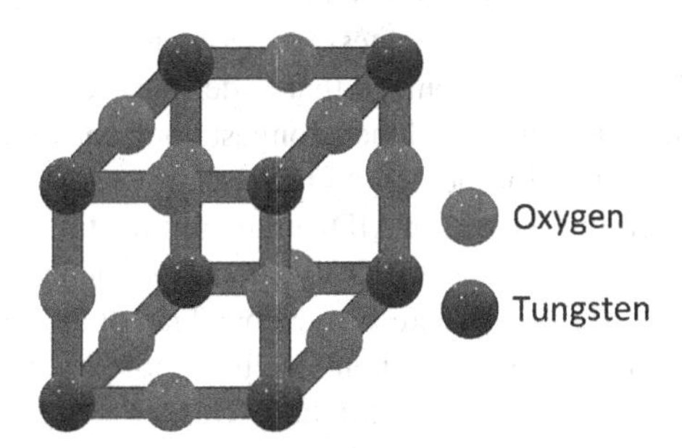

FIGURE 7.1 The ideal cubic structure of WO_3.

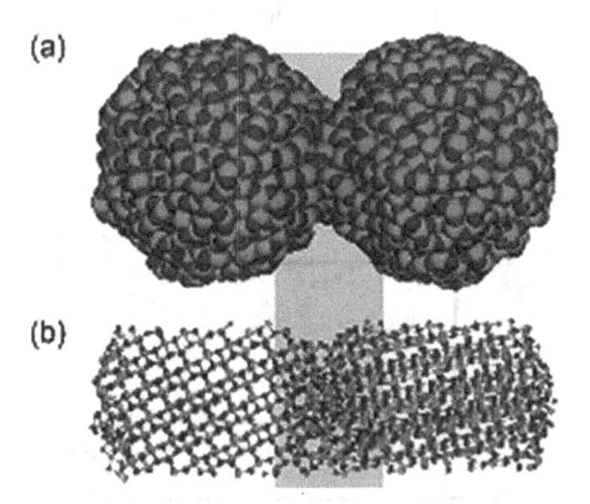

FIGURE 7.2 Molecular dynamic simulation of TiO_2 (a) two 4 nm particles, and (b) cylindrical sample cut at 450°C. (Reprinted (adapted) with permission from Ref. 4. Copyright (2011). American Chemical Society.)

The investigation of the change of pairwise energy over temperature can be analyzed with the assistance of the Nosé–Hoover thermostat algorithm without the Si substrate. Two NO molecules are bonded to the surface of WO_3 over the change of temperature to create N-W bonds [3]. For each simulation, the temperature was initialized at 10 K. Over the change of temperature studied, the bond length of the N-W bond was changed over constant bond energy.

For the first simulation, the bond energy between the N-W bonds was fixed at 13.8 eV [11]. The bond lengths were set at 1.69 Å over temperatures of 10–500 K. The simulation was repeated with bond lengths of 1.67 and 1.65 Å. The simulation was recreated with the bond energy equivalent to 14.5 eV. With the change in temperature over the varying bond lengths, the N-W (1) pairwise energy of the molecule was highest with the bond length of 1.67 Å at the bond energy of 13.8 eV. With bond energy of 14.5 eV, the strongest N-W produced was at a bond length of 1.69 Å. For the N-W (2) under the same conditions, the strongest bond was produced at a bond length of 1.69 Å at 13.8 and 14.5 eV.

The investigation indicated the change of pairwise energy based on the change of bond length and energy. Two NO structures are absorbed on the cubic structure of WO_3 on a silicon substrate. The N-W bonds were created at bond energies of 13.8 and 14.5 eV with bond lengths of 1.69, 1.67, and 1.65 Å. The results show that the second NO structure produced the greatest pairwise energy between the N-W bond at 14.5 eV at 1.69 Å. The results confirm that under bond energies of 14.5 and 13.8 eV, the second NO structure produced the strongest bond energy at 1.69 Å. This does not correspond to the results for the first NO structure.

A case study illustrating the use of MD simulation for tin dioxide (SnO_2) is also presented here [12]. Tin (IV) oxide (SnO_2) is an n-type metal oxide semiconductor material with a tetragonal rutile structure and a bandgap of approximately 3.6 eV (at room temperature) and possesses excellent electron mobility. SnO_2 is often used in technology owing to its stable crystal structure. This structure has properties corresponding to a space-group symmetry of P_{42}/mnm and point symmetry D^{14}_{4h}. As seen in Figure 7.3, each unit cell consists of two cation tin (Sn) six-fold atoms and four anion oxygen (O) three-fold atoms. The lattice constants (a, b, c) for this structure are

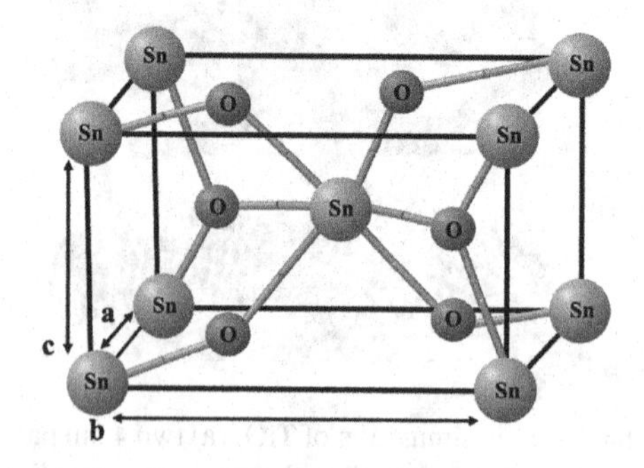

FIGURE 7.3 SnO_2 unit cell. (Reprinted from Ref. 13. Copyright (2009), with permission from Elsevier.)

given to be a = b = 4.7374 Å and c = 3.1864 Å. The tin atoms are located at the vertices of a lattice comprised of six oxygen atoms that form a distorted octahedral.

For SnO_2, within LAMMPS we must use the so-called 'polarizable force field' where the distribution of charges in a molecule interacts with its electrostatic environment. Such a system can be simulated by using an adiabatic core/shell model, which mimics a shell of electrons of an ion by attaching a satellite particle. Therefore, the input file for LAMMPS contains the positions of Sn (core/shell) and O (core/shell). The microcanonical (NVE) ensemble, where the number of atoms (N), the volume of the system (V), and the energy (E) are constant, is employed to determine the lattice parameters of SnO_2. The map-file is generated using the so-called 'Latgen code' with atomic positions, which is required by the fix-phonon command within LAMMPS. This was accomplished using the Buckingham interatomic potential, which is used to simulate the short-range interactions between ions i and j.

$$U_{ij}\left(r_{ij}\right) = A_{ij}e^{\left(-r_{ij}/\rho_{ij}\right)} - C_{ij}/r_{ij}^6 \tag{7.1}$$

For an interatomic distance between i and j of r_{ij}, the first term represents the repulsion interaction, and the second term represents van der Waals interaction (dipole-dipole term). A_{ij}, ρ_{ij} and C_{ij} are constants parameters.

Figure 7.4 shows multiple views of the rutile-like phase SnO_2 supercell with 240 of Sn^{4+} and 240 O^{2-} atomsions.

One can then utilize the Verlet algorithm in tandem with the NPT ensemble, which depends on the initial conditions of particles where the total number of particles, N, the pressure, P, and the temperature, T, are fixed during the simulation, to determine the phonon dispersion modes [15] using LAMMPS over the whole Brillouin zone at different temperatures and thereby identify the transverse optical (TO) modes and longitudinal optical (LO) modes, which in turn can help obtain the Raman-active and infrared-active vibrations for SnO_2. Thus, a simulation of the rutile SnO_2 interaction, along with lattice, bonds, and charge parameters, can facilitate the determination of

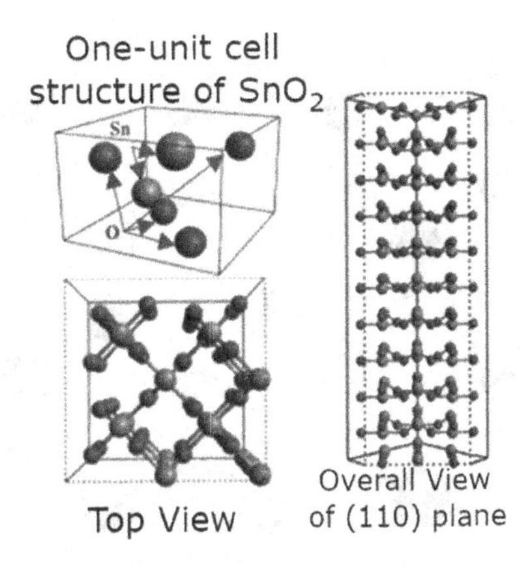

FIGURE 7.4 Different views of an SnO_2 supercell. (Reprinted from Ref. 14. Copyright (2005), with permission from Elsevier.)

the phonon dispersion modes at different temperatures (e.g., in a given range, say, 300–1000 K) affording comparison with experimental vibrational characteristics obtained via Raman and Fourier transform infrared (FT-IR) spectra.

7.3.2 Density Functional Theory

Density functional theory (DFT) is primarily employed to determine the ground state energy of a molecular system composed of a collection of atoms. The calculations can be performed using the Quantum Espresso suite. The Hohenberg-Kohn theorem informs us that the charge density that leads to the lowest energy is the ground state. A good qubit is one with bistable states separated by a wide bandgap of about 6.08 eV. The constraint on bandgap ensures that thermal fluctuations will not cause the system to switch states. We will illustrate the DFT approach for hexagonal boron nitride (hBN), which is a wide bandgap semiconductor [16]. It is known to form a 2D honeycomb pattern, with layers stacked on top of each other, and it possesses strong intralayer bonds, but comparatively weaker interlayer bonds.

The boron vacancy in Figure 7.5 describes a scenario in which a boron atom is absent from its regular position in the hBN periodic lattice. The main motivation for the project was to be able to gain valuable insight into the formation of and properties of defects and whether hBN is a viable qubit candidate for quantum computing. For the boron defect, the DFT calculations were performed on the ferromagnetic (aligned) and antiferromagnetic (antialigned) spin configuration of the material and modelled using periodic boundary conditions. For the ferromagnetic case, a $6 \times 6 \times 1$ supercell was constructed to ensure a realistic defect density for hBN. For the antiferromagnetic case, a $12 \times 6 \times 1$ supercell was used with the spin up and down defects separated by the same distance as with the ferromagnetic arrangement. The most stable configuration is the one with the lowest ground state energy. Strain was applied to the system to visualize and document how the electronic, energetic, and spintronic properties of each configuration change with external strain. As a result of the strain

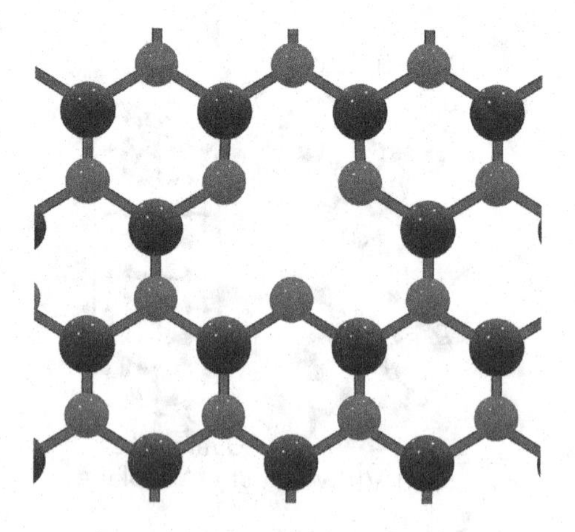

FIGURE 7.5 Boron vacancy in a honeycomb lattice of hexagonal boron nitride (hBN). The B atom is the larger sphere, and the N atom is the smaller one. (Adapted from Ref. 16.)

TABLE 7.1
Strain Dependent Defect-Defect Interactions for Boron Vacancy in hBN [16]

Strain%	Spin	ΔE_{pol}(meV)	ΔE_M(meV)	J(meV)
0	3/2	382.365	23.87	0.88419
1	3/2	432.966	15.36	0.56873
3	3/2	507.114	4.559	0.16887
5	3/2	561.749	2.3164	0.08579
–1	3/2	320.495	25.623	0.94902
–3	1.25/2	259.22	–445.25	–94.987
–5	1.0/2	257.897	–642.74	–214.25

applied, the atomic positions were stretched. Here E_{AFM} and E_{FM} correspond to the formation energy of the antiferromagnetic and ferromagnetic configurations, respectively [16], and E_{LDA} refers to the spin blind total energy. E_{pol} is the polarization energy defined by:

$$E_{pol} = E_{LDA} - E_{FM} \tag{7.2}$$

ΔE_M is the difference in energy between the antiferromagnetic and ferromagnetic cases:

$$\Delta E_M = E_{AFM} - 2E_{FM} \tag{7.3}$$

It was observed that the ferromagnetic defect orientation has a lower energy when the system is not subject to strain. This strainless ferromagnetic configuration exhibits a 3/2 spin. The exchange potential has been determined by:

$$E_M = 12JS^2 \tag{7.4}$$

where J is the exchange coupling and S is the spin.

We note that not only can two spin switches be observed, but in the range of strain studied the sign of E_M changes, which implies that as the system undergoes negative strain, the lowest energy configuration changes from the ferromagnetic case to the antiferromagnetic case.

7.3.3 COMSOL Multiphysics

COMSOL Multiphysics™ is a simulation software used for physics and engineering applications [17]. One of the key features of COMSOL is that it allows the modular increment of physical phenomena; once a model has been developed, the process of adding additional physics becomes relatively straightforward. This allows us to change our model to suit a wide variety of applications and different electric, thermal, and mechanical properties. For example, if a model is developed for the purpose of studying thermal conductivity of a material, it is possible to add physics such as structural mechanics to study additional effects brought out by thermal conductivity. Furthermore, COMSOL has a materials library that can be used to modify the material composition of the object that is being modelled. This makes it possible to easily change the materials being studied for the same object and physics [17]. While the COMSOL library is not all encompassing, it is possible to add custom materials, given that you know the structural properties of the material needed for the model. The combination of the materials library and the different physics modules available, once a model is developed, changing the properties of any of these properties is relatively simple [17]. It is also simple to alter the physical parameters of the model and the formula that govern the physics involved in the model being developed. COMSOL can be used as an effective complement to Raman spectroscopy, as it can determine eigenmodes and eigenfrequencies related to vibrational modes.

COMSOL has been used for modelling the behavior of metal oxide gas sensors with the goal of simulating the ideal conditions these sensors would operate *in situ* [18–20]. Gouthami et al. [19] used COMSOL to design zinc oxide (ZnO) nanowires used for a surface acoustic wave (SAW) sensor, to study the surface deformation caused by the sensing hydrogen (H_2) gas. For this study, COMSOL was used to simulate the structure of a ZnO layer to determine the electrical and mechanical boundary conditions between the model and its surroundings. Figure 7.6 shows the ZnO nanowire SAW sensor COMSOL model.

Work has also been done using COMSOL to model CO gas dispersion in an industrial environment. Figure 7.7 shows a model of CO gas dispersion as a function

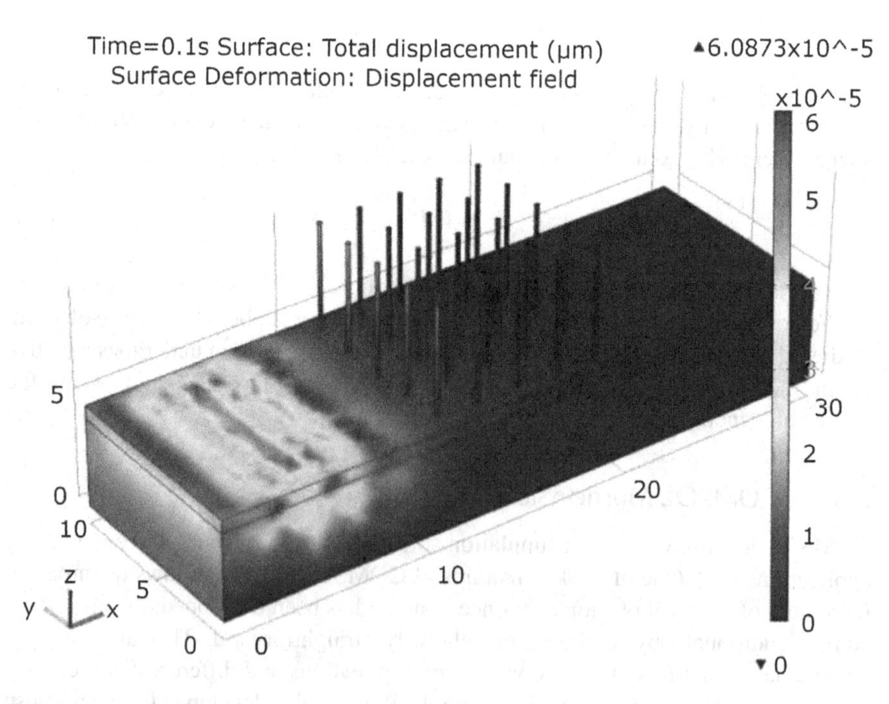

FIGURE 7.6 Simulated result of the ZnO sensor after H_2 exposure [19].

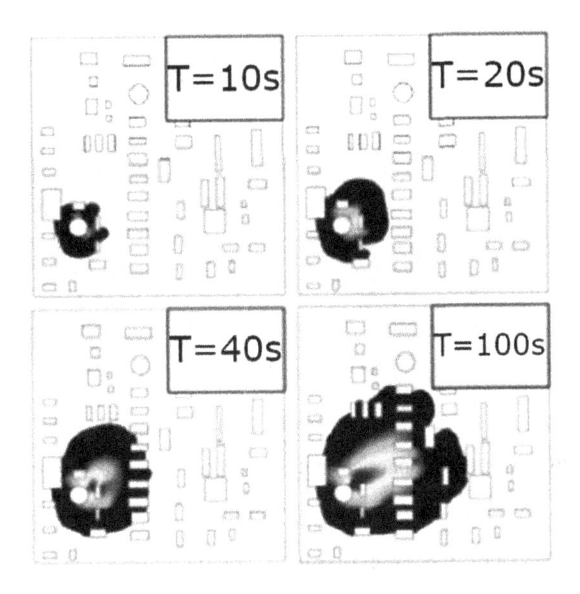

FIGURE 7.7 2D model of CO dispersion as a function of time. (Reprinted from Large Scale Outdoor Flammable & Toxic Gas Dispersion Modelling in Industrial Environments, A. Hallgarth, Copyright (2009), with permission from author.)

of time. As an example of an application of this work, by modelling the gas dispersion over an area, we can determine key locations for the metal oxide gas sensors.

Combining the modelling of gas sensing in metal oxide sensors and the environmental gas dispersion allows modelling of real-life conditions to which these sensors can be applied. By modelling the electrical, mechanical, and thermal parameters of the gas sensor and the behavior of the gas itself, and creating a generalized model, it becomes possible to determine the effectiveness of the gas sensor under a variety of conditions by making use of COMSOL's material library and modular approach.

We will illustrate COMSOL's usage by considering its application for the design and simulation of a heated sample cell for molecular gas sensing applications [21]. As a case study, we will consider its use in designing metal oxide gas sensors (MOGSs). The composition of MOGS can be modified to adjust a wide variety of variables, such as selectivity (what gas the sensor will detect), sensitivity (how sensible the sensor is to the gas), working temperature (the operating temperature of the sensor), among many others. This makes MOGS ideal for detecting a wide variety of gases in different environmental settings. One such material used in the manufacturing of MOGS is tungsten trioxide (WO_3), which often is utilized in the detection of NO_x gases. Car exhausts is a common source for NO_x and can be a respiratory irritant. Monoclinic WO_3 samples 2.5 mm in diameter and 4 μm in thickness were drop-coated and furnaced-fired onto silicon substrates 2.55 × 1.15 × 0.1 cm; this sample was analyzed in a temperature range between 30 and 200°C.

COMSOL Multiphysics was used to model the sample above along with a cell that served as a heating element since the Raman instrument utilized in the experiment did not have inherent heating capabilities. Figure 7.8 illustrates the COMSOL modelling of the heated cell, along with a WO_3 sample deposited on the silicon

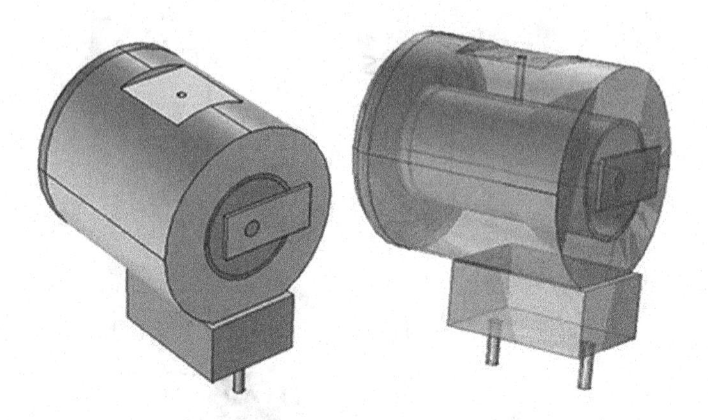

FIGURE 7.8 WO_3 deposited (spherical dot) on silicon substrate (rectangular slide) on the window of the heated cell [21].

substrate and attached to the window of the cell [21]. Figure 7.8 shows how the WO_3 on silicon samples are set up on top of the heating element of the heated cell and, by using the heated cell interface, the samples are heated via conduction to the desired temperature. The heated cell was then placed in front of the laser path in the Raman instrument; this way, it was possible to relate specific temperatures to the outputted Raman graphs.

Using COMSOL, it is possible to assign the material that the different components of our model are composed of. Once the composition of the model has been established, the physics (such as Joule heating) for the model can be defined.

Figure 7.9 shows the surface heatmap over time for a constant electric potential of 0.27209 V (average voltage of the heating cell). This heatmap allows us to see the temperature increase of the heated cell's multiple components and the sample over a period of time.

Table 7.2 shows the results of the temperature heatmap over time of the different heated cell components that were modelled for the simulation. Figure 7.10 shows the arrow surface plot of the current density [3].

Figure 7.11 shows different views of the heated cell surface temperature heatmap for an electric potential that increases over time. Since the electric potential heating occurs from the back of the heated cell, the temperature increases more rapidly from the back and spreads throughout the rest of the cell. It can also be seen that the temperature increase that the sample experiences eventually overtakes the temperature increase from the cell exterior. This way, a basic model of the environment related to electrical potential and temperature mapping to which the sample is subjected when we conduct the temperature experiments on the Ventacon heated cell inside the DXR Raman spectrometer can be created and compared to how it performs in actuality.

The development and addition of more physics toolboxes would allow for more complex simulations and enable modelling of the actual flow of the system, similar to the environment that we have set up for our experiments using the Renishaw Raman Microscope using the 514 nm excitation laser wavelength.

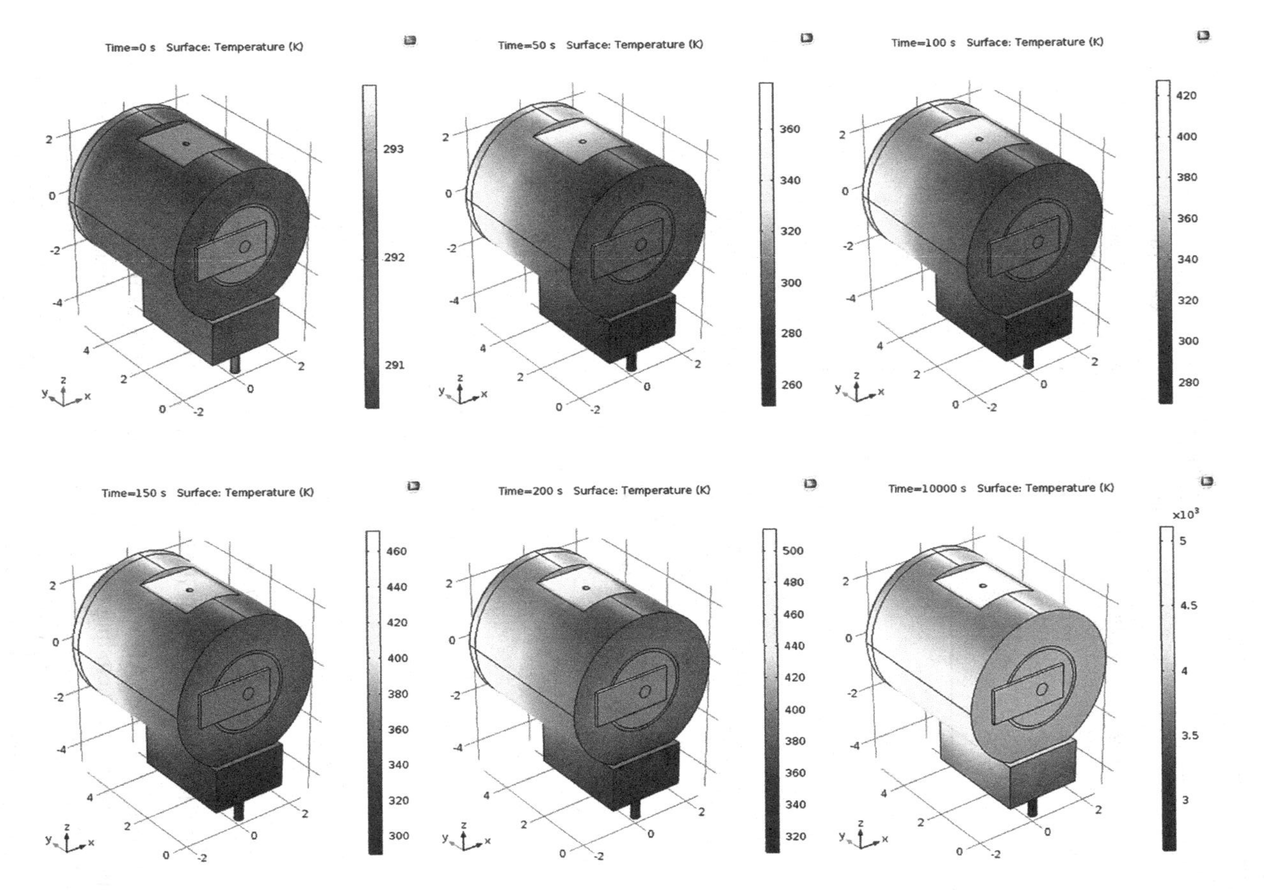

FIGURE 7.9 Front view of the time-dependent temperature plot of the Ventacon heated cell with the WO_3 sample on its surface at t = 0, 50, 100, 150, 200, and 10000 s [3].

TABLE 7.2
Temperature Over Time for the Different Components of the COMSOL Simulation [3]

	Temperature (°C)					
Time (s)	Cell Front	Cell Back	Sample	Felt	Cell Exterior	Cell Bottom
0	20.00	19.89	20.00	20.00	19.86	19.72
50	51.28	96.89	32.87	50.30	48.32	14.54
100	90.93	147.47	66.31	89.67	87.07	43.43
150	132.22	191.64	105.83	130.49	126.11	74.24
200	173.63	233.91	146.61	171.43	165.04	105.40
10000	4776.75	4817.65	4743.52	4761.42	4540.56	3395.23

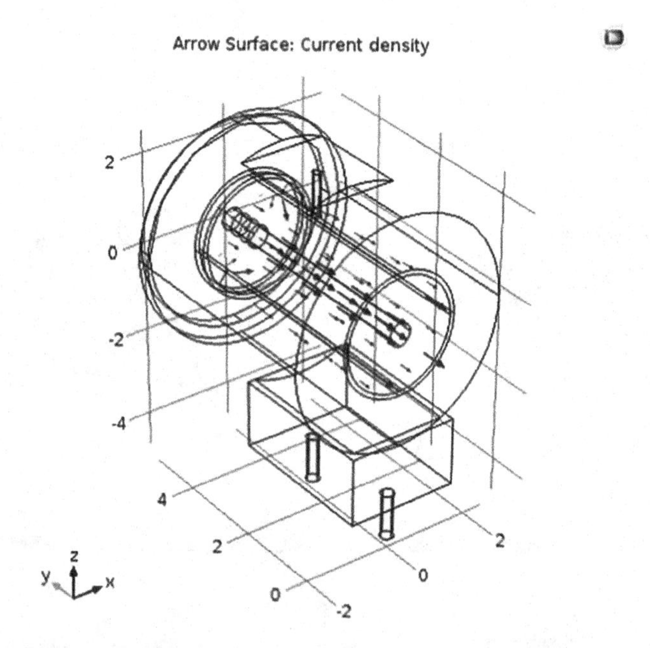

FIGURE 7.10 Current density arrow surface of the heated cell model [3].

7.4 ARTIFICIAL INTELLIGENCE AND MACHINE LEARNING

Artificial intelligence (AI) refers to the simulation of human intelligence by programmed machines, including problem solving, learning, analytical ability, and reasoning [22]. With a specific goal in mind, AI can be used for rational actions and decisions that would lead to the best possible outcome for the scenario. Efficient computer algorithms are a big part of AI and machine learning (ML) [23]. Common applications of AI and ML occur in the financial sector and include large, unusual bank deposits, credit card fraud, and trading of stocks. As a case study, we will investigate the applications of AI and ML in the context of large databases.

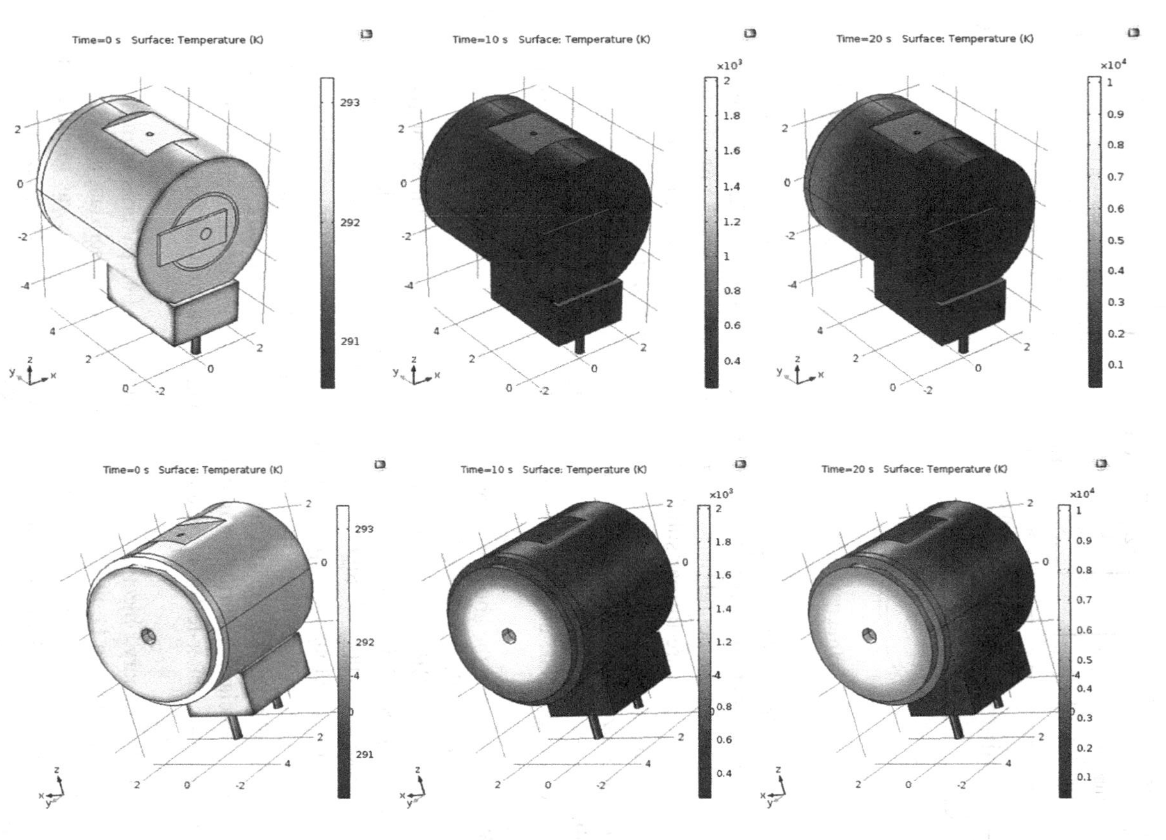

FIGURE 7.11 Front and back view of the Ventacon heated cell under voltage that increases as a function of time [3].

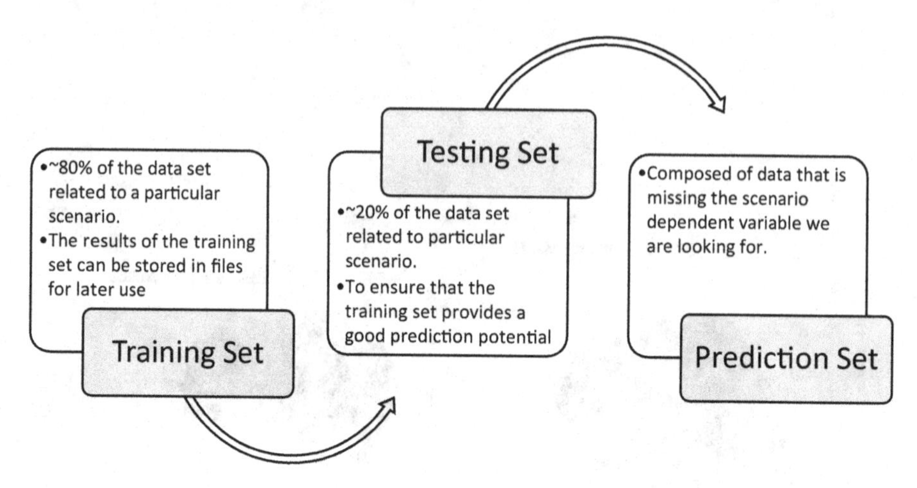

FIGURE 7.12 The neural network data set structural flow diagram. (Adapted from Ref. 24.)

A key concept in developing neural networks revolves around separating the data into three sets: (1) training, (2) test, and (3) prediction [24,25]. The training set data was used to train the neural network and took up most (70–80%) of the total data. The training data were further subdivided by the neural network into training, test, and validation sets, which were used to determine when training results cannot be further optimized. Once the neural network was generated through training, it was applied to the test set data to assess the network's performance, a process called 'cross-validation'. The entire neural network data set structural flow is illustrated in Figure 7.12.

We will illustrate the above with pattern recognition in one large terrorism related database, namely, the profiles of incidents involving chemical, biological, radiological and nuclear (CBRN) events by non-state actors (POICN) database prepared and maintained by START at the University of Maryland, College Park, employing neural networks and other ML algorithms [24,25]. Training and test data subsets were extracted from previously compiled data to develop a variety of models based on ML pattern recognition algorithms that can enable prediction of future terrorist threats with specified percentage errors and uncertainties, and thereby enable the intelligence and homeland security communities to make informed decisions regarding deployment of counter-terrorism resources to effectively thwart terror plots prior to their occurrence.

One of the primary objectives of the project was to find patterns from previously compiled data that can be utilized to predict behavior in newer data by developing a cluster-based algorithm. A central goal of the research project was to utilize neural networks and other ML algorithms to determine missing data from the terrorism-related databases pioneered at the START DHS Center of Excellence (COE) [24,25]. We will focus on POICN for our case study here. For example, data from CBRN attacks in the POICN database from 2000 could be used to teach the algorithm to generate output related to existing data in later years. This, in turn, allowed us to discern patterns based on the multitude of variables used in the POICN data set by looking at links between (1) terrorist attacks based on region, (2) terrorist events and future attacks, and

(3) different terrorist organizations. To implement ML of pattern recognition, several rigorous processes were utilized. Training and test data sets were determined, and earlier data sets were used to predict new data, to develop a ML pattern recognition algorithm and model that could recognize patterns between past and future events.

Akin to pattern recognition, neural networks require the following components: (1) inputs, (2) outputs, (3) training data set, and (4) network topology [24]. Specifically, the inputs are the selected core variables associated with an entry in a particular database (e.g., POICN), and the outputs are patterns recognized by the algorithm based on the inputs and the way the neural network is trained. The training data set is a subset of the database from the earliest entries, which is then used to generate an output that can be compared and matched with test data set entries for validation of the algorithm. A subset of the pattern recognition algorithm developed can then be run to aid in the identification of when specific terrorist groups adopt, for instance, CBRN weapons and, moving forward, which schemes are most probable in posing a security threat to society at large, thereby enabling intelligence and homeland security communities to make informed decisions regarding deployment of counter-terrorism resources to thwart terrorist plots before they occur.

The POICN database demarcates eight different types of CBRN events: use of agent, attempted use, and threat with possession, acquisition of a weapon, acquisition of an agent, attempted acquisition, plot, and protoplot. The eight categories are separated into two broad classes: Type A (seeking a CBRN weapon) and Type B (possessing a CBRN weapon). Both Types A and B are important and to differentiate between the two, an approach is to use the data mining tool of logistic regression analysis, which is one of the numerous data mining tools in the field of knowledge discovery in databases (KDDs) [26].

Figure 7.13 shows the proportions among the 165 sample events that we selected from the 471 total events in the version of the POICN database (available in 2014) that had credibility ratings of 1, 2, or 3 that was used in our logistic regression model of event type prediction. Table 7.3 provides the corresponding count and percentage breakdown of the POICN credibility levels of the events shown in Figure 7.13.

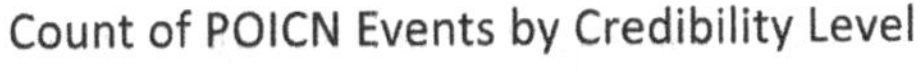

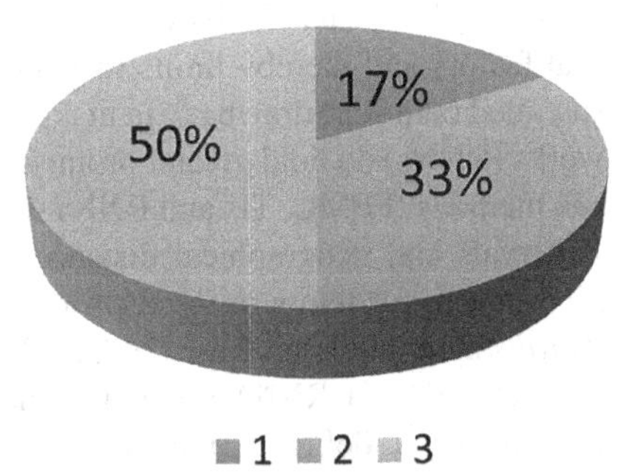

FIGURE 7.13 Pie chart breakdown of credibility levels of POICN events. (Adapted from Ref. 24.)

TABLE 7.3
Breakdown of Credibility Levels of POICN Events

Credibility Level	Count of Credibility	
	Count	Percent
1	53	32
2	21	13
3	91	55
Grand Total	**165**	**100**

Source: Adapted from Ref. 24.

The above technique helped to identify and extract the highest credible sample events from the POICN database to use in an initial logistic regression analysis of CBRN event type prediction. We were able to organize the sample data from the POICN database and ran a linear logistic model regression analysis that has the promise to predict CBRN event type (Type A or Type B) based on the data available in the POICN database.

7.5 APPLICATIONS

7.5.1 Large Databases

Knowledge discovery in databases (KDD) is a technique employed to find useful information from a database. The goal of KDD in databases is to convert the goals and requirements of the end user into a 'data-mining' goal (see Figure 7.14). In the case study that we provided in the previous section, the first step was to determine which database variables we can use for the neural network analysis. Subsequently, we determined potential scenarios for which variables might be useful, excluding those with a high number of unknowns. We carried out neural network tests to determine the effectiveness of specific variables on the misclassification percentages.

Both spatial and temporal contextual information are critical for analyzing user behavior and predicting their subsequent migration and location. A variety of research approaches have been used in this regard, which include the following: (1) Factorizing personalized Markov chain (FPMC) that is based on the strong independence assumption among several factors and thereby limits its application; (2) tensor factorization (TF) has the so-called cold start drawback in making futuristic predictions; (3) recurrent neural networks (RNNs) do hold greater promise as compared to FPMC and TF. However, all three methods (FPMC, TF, and RNN) show limitations in modelling continuous time intervals and geographical distances simultaneously. RNN employs a single transition matrix for temporal prediction that has been applied to word embedding and advertisement clicking. Liu et al. [27] have proposed a spatial temporal-recurrent neural network (ST-RNN) prediction model that incorporates time-specific transition matrices for defining time intervals and distance-specific transition matrices for incorporating geographical distances. The researchers have applied their ST-RNN technique successfully to two large real-world data sets,

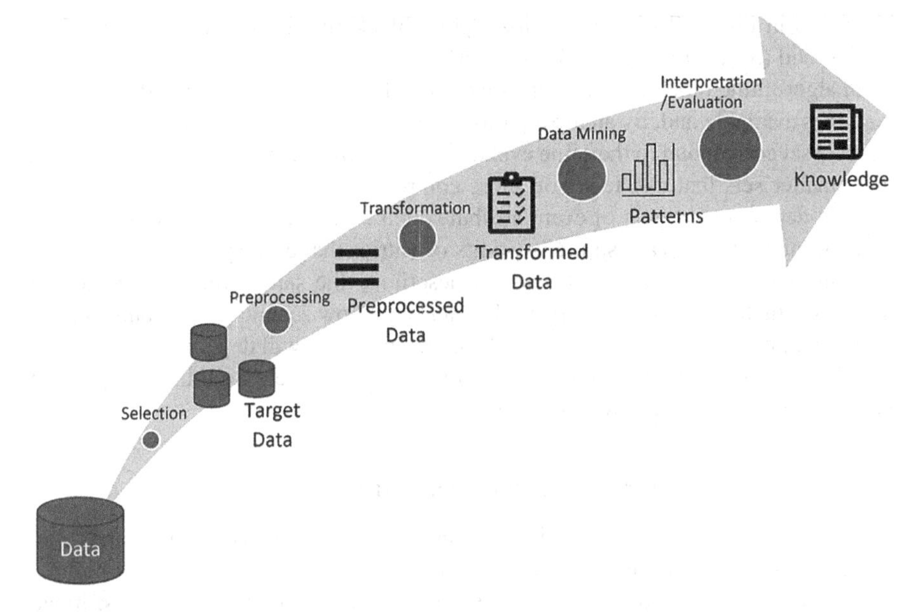

FIGURE 7.14 Process of knowledge discovery in databases.

namely START's Global Terrorism Database (GTD) and the Gowalla data set that is a compendium of check-in history of users.

In addition to the ST-RNN approach, another novel methodology that one can employ utilizes systematic data integration. Figure 7.15 explains the data integration process from data entries from multiple databases [28]. The goal of this process is to compare entries between a variety of data sets that share data on the same event and collect them into a single entity. One must also realize that available records of geo-coordinates, timestamps, and other attributes of events, have a built-in uncertainty or 'fuzziness'. The data integration protocol will be governed by decision-making rules to differentiate between unique and matching events, while including fuzziness in the process of comparing event attributing parameters. Such a protocol is representative of the so-called MELTT methodology [28], which utilizes a three-step process (as

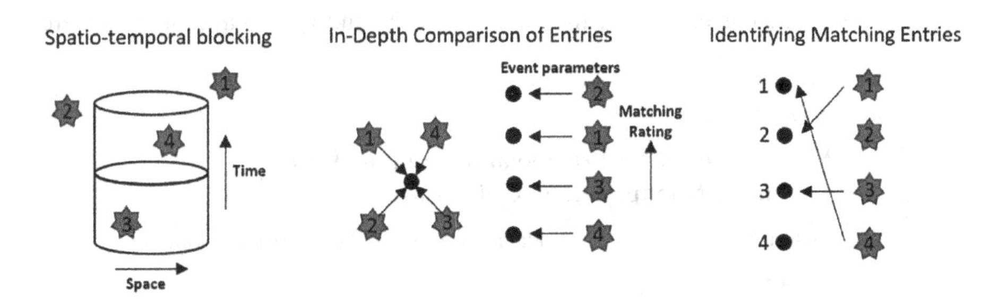

FIGURE 7.15 Data integration process. (Adapted from Ref. 28, Copyright (2018), with permission from Sage Publications under the STM Permission Guidelines.)

illustrated in Figure 7.15): (1) spatiotemporal blocking, (2) detailed comparison of entries, and (3) identifying matching entries.

In algorithmic blocking, the entries would include location and time if that information is available and, by analyzing this information, we can determine if entries in the data set correspond to the same event. Subsequently, we can break down this data into smaller sets that can focus on more compact spatial and temporal event data. Such in-depth comparison of event attributes will be constrained to distinct subsets of adjacent entries. As a result, close pairs of entries that correspond to every event attribute would be identified as possibly describing the same event. The matching index, S, which ranged between 0 and 1, determines how accurately data entries from a particular event match. The value of the matching index will depend on the level of correlation in the data between entries. In turn, this will help establish the pairs of matching entries across different data sets.

7.5.2 Graphitic and Metal Oxide Nanomaterials

Support vector machines (SVMs) implement simple algorithms to analyze and separate data using a hyperplane located in N-dimensional space, where N is the number of parameters [29]. By adding parameters, the data are differentiated and transformed between planes. The transformations that data undergo are referred to as 'kernels'. Applying kernels allows users to discriminate even extreme cases where data might overlap. SVM uses these extreme values of the data set to draw the decision surface. The values that the hyperplane's margin divides on are referred to as 'support vectors'. One of the many positive attributes of SVM is the fact that a huge data set is not necessary to begin the computation. However, result accuracy improves with larger data sets. In addition, the MATLAB® software [30] has a beginning classification learner toolbox available for people to use and facilitates the uploading of a data set in the form of a CSV file into the computer program. These techniques can be applied effectively to graphitic nanomaterials and metal oxides especially for targeted gas sensing applications. As a case study, we will consider pristine and functionalized graphene nanoplatelets (GnPs) [31–33]. Table 7.4 summarizes the materials, and their properties, used for these experiments.

One can prepare the data set from the Raman spectral results of the ammonia (NH_3) and nitrogen (N_2) functionalized GnPs. Twelve trials were performed for each of the functionalized GnPs, while five trials were collected for the pristine GnPs for improved results. For each trial, the D, G, and 2D-band peak wavenumbers were

TABLE 7.4
Materials used for Pristine and Functionalized Graphene Nanoplatelets (GnPs) Case Study [31–33]

Material	Planar Size (μ)	Planar Thickness (nm)
Pristine GnPs	4–12	2–8
Functionalized GnPs	0.3–5	<50
COOH-functionalized GnPs	1–2	3–10

documented, as well as all possible peak-to-peak ratios. The peak wavenumbers used in the modelling and analysis are summarized in Table 7.5, along with the intensity ratios of the D and G bands [31].

In total, there were six parameters: the three peak wavenumbers and their comparative ratios. The ratios were taken into consideration as the number of defects of the graphene should be altered with the functional groups present. In turn, these parameters created a 6D space where the data could be separated. For visualization purposes, Figure 7.16 depicts the layers, or dimensions, that the data filters through and eventually gets quadratically separated.

TABLE 7.5
Average Band Positions of the D, G, and 2D Raman Bands and Intensity Ratios (I_D/I_G) of Pristine and Functionalized GnPs, Along with Experimental Uncertainties

Samples	Band D (cm^{-1})	Band G (cm^{-1})	Band 2D (cm^{-1})	I_D/I_G
Pristine-GnPs	1353.90 ± 3.30	1581.30 ± 1.10	2722.10 ± 2.50	0.25 ± 0.06
GnPs-Nitrogen	1354.80 ± 1.00	1580.60 ± 0.90	2719.60 ± 1.00	0.36 ± 0.03
GnPs-7 wt.% Carboxyl	1353.90 ± 2.00	1581.40 ± 1.70	2721.10 ± 3.70	0.33 ± 0.03
GnPs-35 wt.% Carboxyl	1349.53 ± 5.41	1581.39 ± 1.75	2707.50 ± 10.75	0.93 ± 0.04
GnPs-Ammonia	1355.20 ± 2.20	1582.10 ± 1.40	2721.80 ± 2.30	0.36 ± 0.02
GnPs-Fluorocarbon	1356.00 ± 1.40	1582.30 ± 0.70	2722.70 ± 3.30	0.29 ± 0.02
GnPs-Argon	1356.30 ± 2.20	1582.80 ± 0.20	2723.20 ± 1.40	0.34 ± 0.01
GnPs-Oxygen	1357.20 ± 0.60	1583.20 ± 0.30	2723.80 ± 1.70	0.32 ± 0.01

Source: Reprinted from Ref. 31. Copyright (2021), with permission from Elsevier.

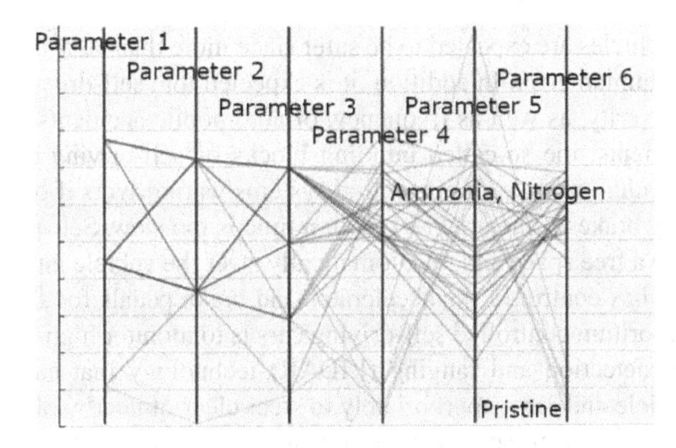

FIGURE 7.16 Network diagram of SVM learning using six parameters to create a 6D quadratically separable data set. As the data move throughout the dimensions, one can see that the pristine and functionalized GnPs become highly separated. Conversely, for the ammonia and nitrogen functionalized nanoplatelets, slight overlapping is still noticeable. However, in this overlapping, a hyperplane surface was created, separating the ammonia and nitrogen functionalized GnPs with good success. (Reprinted from Ref. 31. Copyright (2021), with permission from Elsevier.)

With a working model, it would now be possible to add more functional groups, one at a time sequentially, and expand the model even further with a computer algorithm that would enable differentiation between all six functional groups associated with the GnPs investigated, and eventually be able to characterize those that have made contact with harmful poisonous compounds (e.g., NOx and CO). Large data sets must be prepared to verify the results and build more efficient and versatile models and then be able to develop practical graphitic and metal oxide-based (e.g., WO_3 and SnO_2) toxic gas sensors for field deployment.

7.5.3 Quantum Machine Learning

ML is a powerful tool to detect patterns in data. It is highly likely that quantum systems produce patterns very different from classical systems, and quantum computers can outdo their classical counterparts in performing ML oriented tasks to detect and interpret such patterns. Quantum software is specially designed to build on such advantages by generating quantum states of light and matter and better understand the underlying structure and features of the patterns that get revealed by the algorithms. It has been established that the quantum approach to analyze data representing quantum systems is both more efficient and informative than the classical interpretation of the same data [34]. The use of so-called 'quantum simulators', akin to quantum analog computational machines, is a powerful algorithm-driven approach to investigate the dynamics associated with quantum systems. Quantum-enhanced ML algorithms hold the promise to apply powerful computational resources to design ultrafast next generation quantum processors for a diversity of AI and ML-driven quantum computing applications.

7.5.4 Self-Driving Cars

Autonomous vehicles are expected to be safer since more than 90% of auto accidents happen due to human error. In addition, it is expected that self-driven cars will help decrease the severity, as well as frequency, of automobile accidents [35]. The technology components, the so-called building blocks of self-driving cars, have been around for a while. Front-crash prevention systems warn drivers if obstacles are too close and apply brakes if the driver's reaction time is too slow. Self-park technology helps to size up a free space and to automatically steer the vehicle into the open spot, with the driver has control of the accelerator and brake pedals for safety. The main idea behind algorithm controlled self-driving cars is to monitor high-resolution cameras and light detection and ranging (LIDAR) technology that can track objects around the vehicle and react appropriately to steer clear of nearby objects and avoid a crash. By 2030, it is estimated that the self-driving car market driven opportunities will boom to U.S. $87 billion, with software and ML developers in high demand and sharing a large chunk of that market.

7.6 CONCLUDING REMARKS

This chapter has provided an overview of model development and a diverse variety of simulation and computational techniques, including LAMMPS, DFT, neural network-driven algorithms and COMSOL Multiphysics and their applications to a range of case studies spanning toxic gas sensors, materials-focused chemistry, physics and spectroscopy, and quantum computing. It also covers artificial intelligence and machine learning and quantum computing, with contemporary applications and illuminating case studies involving large databases, artificial intelligence, and machine learning techniques and cybersecurity.

ACKNOWLEDGMENTS

Some of the case studies illustrated in this chapter were facilitated by funding from the National Science Foundation (NSF) via the Research Experiences for Undergraduates (REU) Site in Physics at Howard University (Award Nos. PHY-1358727 and PHY-1659224) and is gratefully acknowledged. These case studies were a direct result of the research performed by the following REU summer interns: Larkin Sayre (2014), Sarah Bartley (2015), Fabiola Diaz (2017), Sean Russell (2018), and Lia Phillips (2019). Larkin, Sarah, Fabiola, and Lia were all mentored by Prof. Prabhakar Misra, while Sean was mentored by Prof. Pratibha Dev at Howard University. In particular, the case study relating to tungsten trioxide was part of Dr. Raul Garcia-Sanchez's Ph.D. dissertation completed under the supervision of Prof. Misra. Additional case studies presented here relating to tin dioxide and graphene nanoplatelets, respectively, draw on the Ph.D. research currently pursued by Hawazin Alghamdi and Olasunbo Farinre under the guidance of Prof. Misra at Howard University. The authors would also like to thank the National Consortium for the Study of Terrorism and Responses to Terrorism (START) at the University of Maryland, College Park, for access to the POICN database and the associated case study presented here, along with the contributions of Dr. Daniel Casimir related to that effort. Funding for the neural network-driven large database project pursued at START was provided by the Department of Homeland Security (DHS) Science and Technology Directorate's Office of University Programs through the Summer Research Team Program for Minority Serving Institutions and is gratefully acknowledged.

REFERENCES

1. Vladimir Mityushev, Wojcieh Nawalaniec, and Natalia Rylko. *Introduction to Mathematical Modeling and Computer Simulations*. CRC Press, Boca Raton, FL. ISBN-13: 978-1-138-19765-7, 2018.
2. Large-scale Atomic/Molecular Massively Parallel Simulator (LAMMPS) Molecular Dynamics Simulator. https://lammps.sandia.gov/
3. Raul Garcia-Sanchez, *Characterization of Metal Oxide Gas Sensor Materials Using Raman Spectroscopy and Computer Simulations*, Ph.D. Dissertation, Howard University, Washington, DC, 2016.
4. S.J. Konezny, C. Richter, R.C. Snoeberger III, A.R. Parent, G.W. Brudvig, C.A. Schmuttenmaer, and V.S. Batista. Fluctuation-Induced Tunneling Conductivity in Nanoporous TiO2 Thin Films. *Journal of Physical Chemistry Letters* 2, 1931; 2011. dx.doi.org/10.1021/jz200853v
5. Robert Bondi. Atomistic Modeling of Memristive Switching Mechanisms in Transition Metal Oxides. Sandia National Laboratories, *LAMMPS Users' Workshop and Symposium*, August 2011.

6. Y.F. Sun, S.B. Liu, F.L. Meng, J.Y. Liu, Z. Jin, L.T. Kong, and J.H. Liu. Metal oxide nanostructures and their gas sensing properties: A review. *Sensors*, 12, 2610-2631, 2012. doi:10.3390/s120302610
7. B. Chwieroth, B.R. Patton, and Y. Wang. Conduction and gas surface reaction modeling in metal oxide gas sensors. *Journal of Electroceramics*, 6(1), 27–41, 2001.
8. M. Boulova, A. Gaskov, and G. Lucazeau. Tungsten oxide reactivity versus CH_4, CO and NO_2 molecules studied by Raman spectroscopy. *Sens. Actuators, B* 81, 99, 2001.
9. X. An et al. WO_3 nanorods/graphene nanocomposites for high-efficiency visible-light-driven photocatalysis and NO_2 gas sensing. *Journal of Materials Chemistry*, 22, 8525, 2012. doi:10.1039/c2jm16709c
10. J. Qin et al. Graphene-wrapped WO_3 nanoparticles with improved performances in electrical conductivity and gas sensing properties, *Journal of Materials Chemistry*, 21, 17167, 2011. doi:10.1039/c1jm12692j
11. Sarah Bartley. *Gas sensing properties of the adsorption of NO on WO3 cubic structures of different bond lengths*. Final report for the REU Program in Physics at Howard University, Washington, DC, 2015.
12. Hawazin Alghamdi, Benjamin Concepcion, Shankar Baliga, and Prabhakar Misra. Synthesis, Spectroscopic Characterization and Applications of Tin Dioxide. In N.M. Mubarak, R. Walvekar, N. Arshid, and M. Khalid (eds.) *Contemporary Nanomaterials in Material Engineering Applications, Engineering Materials*. Springer Nature, Switzerland AG, 2021. https://doi.org/10.1007/978-3-030-62761-4_11.
13. P. Armstrong, C. Knieke, M. Mackovic, G. Frank, A. Hartmaier, M. Göken, and W. Peukerta. Microstructural evolution during deformation of tin dioxide nanoparticles in a comminution process. *Acta Materialia*, 57(10), June 2009, 3060–3071.
14. Matthias Batzill and Ulrike Diebold, The surface and materials science of tin oxide. *Progress in Surface Science*, 79(2–4), 2005, 47–154. ISSN 0079-6816. https://doi.org/10.1016/j.progsurf.2005.09.002.
15. A. Schleife, J.B. Varley, F. Fuchs, C. Rödl, F. Bechstedt, P. Rinke, and C.G. Van de Walle. Tin dioxide from first principles: Quasiparticle electronic states and optical properties. *Physical Review B*, *83*(3), 035116, 2011.
16. Evan Folk and Pratibha Dev. *Doping Boron Nitride*. Final report for the REU Program in Physics at Howard University, Washington, DC, 2018.
17. COMSOL Multiphysics Simulation Software. https://www.comsol.com/
18. A. Auge, A. Weddemann, B. Vogel, F. Wittbracht, and A. Hütten. *Oxidation of metallic nanoparticles. Proceedings of the COMSOL Conference*, Milan, Italy, 2009.
19. N. Gouthami, D. Parthiban, M. Alagappan, and G. Anju. *Design and simulation of 3D ZnO nanowire-based gas sensor for conductivity studies. Proceedings of the COMSOL Conference in Bangalore*, India, 2011.
20. A. Hallgarth, A. Zayer, A. Gatward, and J. Davies. *Large scale outdoor flammable and toxic gas dispersion modeling in industrial environments. Proceedings of the COMSOL Conference in Milan*, Italy, 2009.
21. Larkin Sayre. *Raman Spectroscopy and COMSOL Multiphysics Simulation Studies of Tungsten Oxide (WO3) as a Potential Metal-Oxide Gas Sensor (MOGS). REU in Physics Program at Howard University*, 2014.
22. Weiyu Wang and Keng Siau. Artificial intelligence, machine learning, automation, robotics, future of work and future of humanity: A review and research agenda. *Journal of Database Management*, 30(1), 2019. doi:10.4018/JDM2019010104
23. Oswald Campesato. *Artificial Intelligence, Machine Learning, and Deep Learning*, Mercury Learning and Information LLC, Dulles, VA, 2020. ISBN: 978-1-68392-467-8.

24. Raul Garcia-Sanchez, Daniel Casimir, and Prabhakar Misra. *Innovative Algorithm and Database Development Relevant to Counterterrorism and Homeland Security Efforts at START, National Consortium for the Study of Terrorism and Responses to Terrorism (START) Report*, University of Maryland, College Park, MD, August 2014.
25. Prabhakar Misra, Raul Garcia-Sanchez, and Daniel Casimir. Development and Optimization of Machine Learning Algorithms and Models of Relevance to START Databases, Report to the Office of University Programs, Science & Technology Directorate, U.S. Department of Homeland Security, National Consortium for the Study of Terrorism and Responses to Terrorism (START) Report, University of Maryland, College Park, MD, April 2016.
26. U. Fayyad, G. Piatetsky-Shapiro, and P. Smyth. From data mining to knowledge discovery in databases. *AI Magazine*, 17(3), 37, 1996. https://doi.org/10.1609/aimag.v17i3.1230
27. Q. Liu, S. Wu, L. Wang, and T. Tan. Predicting the Next Location: A Recurrent Model with Spatial and Temporal Contexts. *Proceedings of the Thirteenth AAAI Conference on Artificial Intelligence*, AAAI-16, 2016, pp. 194–200.
28. Karsten Donnay, Eric T. Dunford, Erin C. McGrath, David Backer, and David E. Cunningham. Integrating conflict event data. *Journal of Conflict Resolution,* 63(5), 1337–1364, 2018. https://doi.org/10.1177/0022002718777050
29. S. Patel. Chapter 2: SVM (Support Vector Machine) -Theory. *Machine Learning*, 101, 2017.
30. MATLAB software. https://www.mathworks.com/products/matlab.html
31. Daniel Casimir, Raul Garcia-Sanchez, Olasunbo Farinre, Lia Phillips, and Prabhakar Misra, Raman Spectroscopy and Molecular Dynamics Simulation Studies of Graphitic Nanomaterials. In *Modeling, Characterization and Production of Nanomaterials: Electronics, Photonics and Energy Applications,* 2nd ed., Vinod Tewary and Yong Zhang (eds.), Elsevier Science, Woodhead Publishing Series in Electronic and Optical Materials Series, ISBN-10: 0128199059, ISBN-13: 9780128199053, New York, NY, 2021.
32. Daniel Casimir, Iman Ahmed, Raul Garcia-Sanchez, Prabhakar Misra, and Fabiola Diaz, *Raman Spectroscopy of Graphitic Nanomaterials, Chapter in Raman Spectroscopy*. Gustavo M. do Nascimento (ed.), InTechOpen, London, England, 2017. http://dx.doi.org/10.5772/intechopen.72769
33. Daniel Casimir, Hawazin Alghamdi, Iman Y. Ahmed, Raul Garcia-Sanchez, and Prabhakar Misra, *Raman Spectroscopy of Graphene, Graphite and Graphene Nanoplatelets, Chapter in 2D Materials*. Chatchawal Wongchoosuk (ed), InTechOpen, London, England, 2019. http://dx.doi.org/10.5772/intechopen.84527.
34. Maria Schuld, Ilya Sinayskiy, and Francesco Petruccione. An introduction to quantum machine learning. *arXiv:1409.3097v1 [quant-phy]* 10 Sep 2014.
35. P. A. Hancock, Illah Nourbakhsh, and Jack Stewart. On the Future of Transportation in an Era of Automated and Autonomous Vehicles. *Proceedings of the National Academy of Sciences*, 116 (16) 7684–7691, 2019. doi:10.1073/pnas.1805770115.

Appendix

A.1 FOURIER TRANSFORM

Fourier transform is an important tool in signal processing and communication to study the spectrum of various signals. Using Fourier transform, a signal, $x(t)$, in time domain can be represented in frequency domain as

$$X(\omega) = \int_{-\infty}^{\infty} x(t) e^{-j\omega t} dt \tag{A.1}$$

The Fourier transform pair is represented as

$$x(t) \rightleftarrows X(\omega) \tag{A.2}$$

where $x(t)$ is the function/signal in time domain, and $X(\omega)$ is the Fourier transform of $x(t)$.

A.1.1 Fourier Transform of Some Common Functions

The Fourier transform pairs for some common functions/signals are listed below:

1. $$1 \rightleftarrows 2\pi\delta(\omega) \tag{A.3}$$

2. $$\delta(t) \rightleftarrows 1 \tag{A.4}$$

3. $$e^{-at}u(t) \rightleftarrows \frac{1}{a + j\omega} \qquad a > 0 \tag{A.5}$$

4. $$e^{at}u(-t) \rightleftarrows \frac{1}{a + j\omega} \qquad a > 0 \tag{A.6}$$

5. $$u(t) \rightleftarrows \pi\delta(\omega) + \frac{1}{j\omega} \tag{A.7}$$

6. $$\cos(\omega_0 t) \rightleftarrows \pi\left[\delta(\omega - \omega_0) + \delta(\omega + \omega_0)\right] \tag{A.8}$$

7. $$\sin(\omega_0 t) \rightleftarrows j\pi\left[\delta(\omega + \omega_0) - \delta(\omega - \omega_0)\right] \tag{A.9}$$

8. $$\text{rect}\left(\frac{t}{\tau}\right) \rightleftarrows \tau\,\text{sinc}\left(\frac{\omega\tau}{2}\right) \tag{A.10}$$

   where $\text{rect}\left(\frac{t}{\tau}\right)$ is rectangular pulse of width τ defined as

$$\text{rect}\left(\frac{t}{\tau}\right) = \begin{cases} 1, & |t| < \frac{\tau}{2} \\ \frac{1}{2}, & |t| = \frac{\tau}{2} \\ 0, & |t| > \frac{\tau}{2} \end{cases} \tag{A.11}$$

9. $$\text{sgn}(t) \rightleftarrows \frac{2}{j\omega} \tag{A.12}$$

where sgn (t) is sign function (also referred to as 'signum function') defined as

$$\text{sgn}(t) = \begin{cases} 1, & |t| > 0 \\ 0, & |t| = 0 \\ -1, & |t| < 0 \end{cases} \tag{A.13}$$

10. $$te^{-at}u(t) \rightleftarrows \frac{1}{(a+j\omega)^2} \qquad a > 0 \tag{A.14}$$

11. $$t^n e^{-at}u(t) \rightleftarrows \frac{n!}{(a+j\omega)n^{+1}} \qquad a > 0 \tag{A.15}$$

It can be observed that the Fourier transform of most functions is complex or imaginary. So, the Fourier transform can be represented as follows:

- Magnitude spectrum: The plot of magnitude of the Fourier transform with frequency is known as 'magnitude spectrum'.
- Phase spectrum: The plot of phase of the Fourier transform with frequency is known as 'phase spectrum'.

Non-zero values of magnitude spectrum signify the presence of corresponding frequency component in the signal, and the non-zero phase spectrum signifies the phase of corresponding frequency component.

A.1.2 Properties of Fourier Transform

For the Fourier transform pair, as shown in expression (A.2), some properties are listed in Table A.1.

From the properties summarized in the above table, the following remarks are important to note:

- From the property of scalar multiplication and superposition, it can be shown that the Fourier transform is a linear transform.

- The property of multiplication can be used to obtain the frequency spectrum of AM signals.
- The property of convolution is useful in analyzing linear time invariant systems in frequency domain.

A.2 ORTHOGONAL AND ORTHONORMAL SIGNALS

Two signals are called orthogonal if the integral of their product over a given duration is zero. Mathematically, the signals $x(t)$ and $y(t)$ are called orthogonal if they satisfy the following condition:

$$\int_T x(t)y(t)dt = 0 \tag{A.16}$$

Further, the signals are called orthonormal if they satisfy the above condition and are normalized to unit energy, i.e.,

$$\int_T |x(t)|^2 dt = \int_T |y(t)|^2 dt = 1 \tag{A.17}$$

TABLE A.1
Table to Summarize the Properties of Fourier Transform

Property	Signal in Time Domain	Fourier Transform
Scalar multiplication	$kx(t)$	$kX(\omega)$
Superposition	$x_1(t) + x_2(t)$	$X_1(\omega) + X_2(\omega)$
Duality	$X(t)$	$2\pi x(-\omega)$
Time shifting	$x(t - T)$	$e^{-j\omega T} X(\omega)$
Time scaling	$x(at)$	$\frac{1}{\lvert a \rvert} X\left(\frac{\omega}{a}\right)$
Frequency shifting	$e^{-j\omega_0 T}x(t)$	$X(\omega - \omega_0)$
Multiplication	$x_1(t)\, x_2(t)$	$\frac{1}{2\pi} X_1(\omega) * X_2(\omega)$
Convolution	$x_1(t) * x_2(t)$	$X_1(\omega)\, X_2(\omega)$
Differentiation	$\frac{d^n x(t)}{dt^n}$	$(j\omega)^n X(\omega)$
Integration	$\int_{-\infty}^{t} x(\tau)d\tau$	$\frac{1}{j\omega} X(\omega) + \pi X(0)\delta(\omega)$

- The property of multiplication [illegible] to be used to obtain the frequency spectrum of AM s[illegible]
- The property of convolution [illegible] in [illegible] time [illegible] system[illegible] in frequency domain.

[illegible] ORTHOGONAL [illegible] ORTHO[illegible] SIGNALS

[illegible] signals [illegible] of the [illegible] is zero. Mathe[illegible] the following [illegible]

[illegible]

Index

A

B

C

Printed in the United States
by Baker & Taylor Publisher Services